FOOD ANALYSIS AND PRESERVATION

Current Research Topics

FOOD ANALYSIS AND PRESERVATION

Current Research Topics

Edited By

Michael G. Kontominas, PhD

Apple Academic Press

TORONTO NEW JERSEY

Apple Academic Press Inc.
3333 Mistwell Crescent
Oakville, ON L6L 0A2
Canada

Apple Academic Press Inc.
1613 Beaver Dam Road, Suite # 104
Point Pleasant, NJ 08742
USA

Exclusive worldwide distribution by CRC Press, a Taylor & Francis Group

International Standard Book Number: 978-1-926895-07-9 (Hardback)

Library of Congress Control Number: 2012935666

Library and Archives Canada Cataloguing in Publication

Food analysis and preservation: current research topics/edited by Michael G. Kontominas.

Includes bibliographical references and index.
ISBN 978-1-926895-07-9
1. Food–Analysis. 2. Food--Preservation. 3. Food–Microbiology. I. Kontominas, Michael G.

TX541.F66 2012 664 C2011-908708-1

DEDICATION

Dedicated to father George who taught me that hard work can be both fun and fulfilling and to son George who will hopefully embrace the same ideas.

Contents

List of Contributors

D. Alevras
Laboratory of Food Chemistry and Technology, Department of Chemistry, University of Ioannina, Ioannina 45110, Greece.

Anastasia V. Badeka
Laboratory of Food Chemistry, Department of Chemistry, University of Ioannina, 45110 Ioannina, Greece.

E.Chouliara
Laboratory of Food Chemistry and Microbiology, Department of Chemistry, University of Ioannina, Ioannina 45110, Greece.

N. Chounou
Laboratory of Food Chemistry and Food Microbiology, Department of Chemistry, University of Ioannina, Ioannina 45110, Greece.

Dimitrios Goergantelis
Laboratory of Food Chemistry and Food Microbiology, Department of Chemistry, University of Ioannina, Ioannina 45110, Greece.

Ioannis Karabagias
Laboratory of Food Chemistry and Technology, Department of Chemistry, University of Ioannina, Ioannina 45110, Greece.

E.S. Karakosta
Laboratory of Food Chemistry, Department of Chemistry, University of Ioannina, 45110-Ioannina, Greece.

K. Karakostas
Section of Statistics, Department of Mathematics, University of Ioannnina,Ioannina 45110, Greece.

Panagiota Katikou
National Reference Laboratory of Marine Biotoxins, Institute of Food Hygiene, Ministry of Rural Development and Foods, Limnou 3A, 54627 Thessaloniki, Greece.

Michael G. Kontominas
Laboratory of Food Chemistry, Department of Chemistry, University of Ioannina, 45110 Ioannina, Greece.

Artemis P. Louppis
Laboratory of Food Chemistry, Department of Chemistry, University of Ioannina, 45110 Ioannina, Greece.

S. F. Mexis
Laboratory of Food Chemistry and Food Microbiology, Department of Chemistry, University of Ioannina, Ioannina 45110, Greece.

E. Pagiataki
Laboratory of Food Chemistry and Technology, Department of Chemistry, University of Ioannina, Ioannina 45110, Greece.

Evangelos K. Paleologos
General Chemical State Laboratory, Ioannina Division, Dompoli 30, 45332, Ioannina, Greece.

Nikolaos Pournis
Laboratory of Food Chemistry and Technology, Department of Chemistry, University of Ioannina, Ioannina 45110, Greece.

M. Revi
Laboratory of Food Chemistry and Technology, Department of Chemistry, University of Ioannina, Ioannina 45110 – Greec.

K.A. Riganakos
Laboratory of Food Chemistry, Department of Chemistry, University of Ioannina, 45110-Ioannina, Greece.

A. Sacco
Laboratory of Physical Chemistry, University of Bari, Bari 70126, Italy.

M. Tasioula-Margari
Laboratory of Food Chemistry and Technology, Department of Chemistry, University of Ioannina, Ioannina 45110, Greece.

Panagiota D. Zygoura
Laboratory of Food Chemistry and Technology, Department of Chemistry, University of Ioannina, Greece.

List of Abbreviations

AA	Ascorbic acid
ADAM	9-Athryldiazomethane
ANOVA	Analysis of variance
ATBC	Acetyl tributyl citrate
BOD	Biological oxygen demand
BPW	Buffered peptone water
CFC	Cetrimide, fucidine and cephaloridine
COD	Chemical oxygen demand
DG	Digestive glands
DOA	Dioctyl adipate
DSP	Diarrhetic shellfish toxin
DTX-1	Dinophysistoxin-1
DTX-2	Dinophysistoxin-2
DTX-3	Dinophysistoxin-3
DTX's	Dinophysistoxins
ECN	Equivalent carbon number
EOs	Essential oils
EP	Edible parts
ESI	Electrospray ionization
EU	European Union
EVA	Ethylene vinyl acetate
EVOOs	Extra virgin olive oils
FID	Flame ionization detector
FLD	Fluorometric detection
GC-MS	Gas Chromatography–Mass Spectrometry
GRAS	Generally recognized as safe
GSRT	General Secretariat of Research and Technology
GYM	Gymnodimine
HDPE	High density polyethylene
HP	Hepatopancreas
HPLC-FLD	High performance liquid chromatography with fluorometric detection
IS	Internal standard
LAB	Lactic acid bacteria
LC-MS/MS	Liquid chromatography coupled with tandem mass spectrometry

LC-ISP-MS	Liquid chromatography-mass spectrometry
LDA	Linear discriminant analysis
LDPE	Low density polyethylene
LDPE/PA/LDPE	Low density polyethylene/polyamide/low density polyethylene
LOD	Limit of Detection
LOQ	Limit of Quantification
LSD	Least significance difference
MDA	Malondialdehyde
MRS	Man rogosa sharpe agar
MU	Mouse units
OA	Okadaic acid
OEO	Oregano essential oils
O_3	Ozone
PC	Polycarbonate
PE	Polyethylene
PET	Polyethylene terephthalate
PVDC/PVC	Plasticized polyvinylidene chloride/polyvinyl chloride
RCM	Reinforst clostridium medium
RH	Relative humidity
RP-HPLC	Reversed-phase high performance liquid chromatography
RT	Retention time
SCF	Scientific Committee on Food
SPE	Solid Phase Extraction
SPME-GC/MS	Solid phase microextration-gas chromatography
SPME	Solidphase microextraction
SSO	Specific spoilage organisms
TBA	Thiobarbituric acid test
TDI	Tolerable daily intake
TEO	Thyme essential oils
TG	Triglyceride
TVC	Total viable counts
YTX	Yessotoxins

Preface

The book in hand focuses on specific studies in Food Analysis and Preservation carried out in the Laboratory of Food Chemistry and Technology, Department of Chemistry, University of Ioannina, Greece over the past five years. Food Analysis and Preservation aim at the same basic target which is consumer protection. Foods are being processed to preserve quality and prevent spoilage caused by physical, chemical and mostly microbiological agents. In this sense, microbiology is inherently related to food preservation. In turn, Food Analysis provides invaluable information regarding food substrates, toxicology, nutritional content, microbiology etc. As opposed to traditional wet chemistry analysis, instrumental food analysis has developed as an essential tool for the food chemist. Progress in analytical chemistry enables determination of trace amounts of natural constituents as well as contaminants which seemed to be impossible a few years ago.

Food Processing on the other hand, based on contemporary consumers' demands has turned to innovative methods of preservation, resulting to nutritionally healthier, more natural, less heavily processed and microbiologically safe foods of high sensorial quality. The experimental studies included in this book focus on modern analytical techniques including GC, HPLC, GC/MS, LC/MS used to determine either natural constituents of foods or food contaminants in minimally processed foods. Preservation methods include essential oils, bacteriocins, chitosan, packaging and ozonation.

This book emphasizing the above interrelationships between Food Analysis, Food Processing/Preservation and Food Microbiology will be valuable to Food Scientists around the world engaging in respective research fields.

I would like to thank all my colleagues and graduate students for their "contributions", their patience and their cooperation during the preparation of this volume.

— Michael G. Kontominas, PhD

Chapter 1

Determination of DSP Toxins in Mussels and Decontamination using Ozonation

Artemis P. Louppis, Anastasia V. Badeka, Panagiota Katikou, Dimitrios Goergantelis, Evangelos K. Paleologos, and Michael G. Kontominas

INTRODUCTION

In the first part of the study an approach involving both chemical and biological methods was undertaken for the detection and quantification of the marine toxins okadaic acid (OA), dinophysistoxin-1 (DTX-1) and their respective esters in mussels from different sampling sites in Greece during the period 2006–2007. Samples were analyzed by means of (a) high performance liquid chromatography with fluorometric detection (HPLC-FLD) using 9-athryldiazomethane (ADAM) as pre-column derivatization reagent, (b) liquid chromatography coupled with tandem mass spectrometry (LC-MS/MS), and (c) the mouse bioassay (MBA). Free OA and DTX-1 were determined by both HPLC-FLD and LC-MS/MS, while their respective esters were determined only by LC-MS/MS after alkaline hydrolysis of the samples. The detection limit (LOD) and quantification limit (LOQ) of the HPLC-FLD method were 0.015 mg/g HP and 0.050 mg/g HP, respectively, for OA. The LOD and LOQ of the LC-MS/MS method were 0.045 mg/g HP and 0.135 mg/g HP, respectively, for OA. Comparison of results between the two analytical methods showed excellent agreement (100%), while both HPLC-FLD and LC-MS/MS methods showed an agreement of 97.1% compared to the MBA.

In the second part of the study an attempt was made to reduce toxin content of contaminated shucked mussels collected during the diarrhetic shellfish toxin (DSP) episodes of 2007 and 2009 in Greece using ozone treatment. Ozonation resulted in toxin reduction in the range of 6–100%, 25–83%, and 21–66% for free OA, OA esters and total OA respectively. Percent reduction of OA content was substantially higher in homogenized mussel tissue as compared to whole shucked mussels. Upon optimization, ozonation shows promising potential for commercial DSP detoxification purposes.

The DSP is a gastrointestinal disease caused by ingestion of shellfish contaminated by OA and/or dinophysistoxins (DTX's) produced by marine dinoflagellates belonging to the genera *Dinophysis* and *Prorocentrum*. Shellfish, such as mussels, filter approximately 20 l/hr of water. During algal blooms water may contain up to several million algae per liter. Although not all algae cells produce toxins, it is estimated that a significant accumulation of toxins will occur in mussels (Christian and Louckas, 2008). The OA and to a lesser extent, its methylated analogue DTX-1 have been identified as being responsible for most DSP outbreaks in Greece (Mouratidou et al., 2004;

Prassopoulou et al., 2009). Besides OA and DTX-1, DTX-2 has been implicated as an important DSP toxin in Irish (Carmody et al., 1996), Galician (Gago-Martinez et al., 1996) and Portugese (Vale and Sampayo, 2002) shellfish. Additionally, DTX-3, a complex mixture of 7-O-acyl derivatives of DTX-1 ranging from tetradecanoic acid (14:0) to docosahexaenoic acid (C_{22}:6, ω 3), was the main diarrhetic toxin found in scallops in Japan (Yasumoto et al., 1985). Human consumption of shellfish containing DSP toxins exceeding certain levels results in diarrhea, nausea, vomiting, and abdominal pain. Symptoms begin 30 min to a few hours after ingestion while complete recovery occurs within 3 days (Yasumoto et al., 1978). The minimum doses of OA and DTX-1 necessary to induce above symptoms in adults have been estimated to be 40 and 36 μg respectively (Hamano et al., 1986). According to relevant data of the National Reference Laboratory on Marine Biotoxins of Greece (Thessaloniki, Greece), toxins of the OA group have been identified as being responsible for the majority of DSP outbreaks during the past 10 years in Greece (Prassopoulou et al., 2009).

At present the MBA still remains the reference method for the detection of lipophilic algal toxins (EU Regulation 2074/2005) and all analytical methods used for this purpose have to be evaluated against bioassays. The MBA was first developed by Yasumoto et al. (1978). Toxicity in the MBA method is expressed in mouse units (MU). One MU is defined as the minimum quantity of toxin capable of killing a mouse of 20 g in weight within 24 hr after intraperitoneal injection (Yasumoto et al., 1980). This is equivalent to approximately 4 μg of OA for the ddY mouse strain in Japan but may vary somewhat with the strain (Fernandez et al., 2003).

Such bioassays, however, reveal only the total toxicity of a sample providing no indication of each individual algal toxin involved in a given outbreak. A further important argument against bioassays is the growing ethical concerns against the use of laboratory animals (Fernandez et al., 2003). In addition, the MBA is time consuming and may give false positive results because of interferences by other toxins, such as saxitoxin (Fernandez et al., 2003), Gymnodimine (GYM) and Spirolides (SPXs) (Suzuki et al., 2005) or by fatty acids (Suzuki et al., 1996; Tagaki et al., 1984). These problems have led to the introduction of HPLC (Lee et al., 1987, Mouratidou et al., 2004, Prassopoulou et al., 2009, Vale and Sampayo 1999) and more recently to LC-MS based methods (Christian and Luckas, 2008, Fux et al., 2007, Gerssen et al., 2009, Pleasance et al., 1992, Suzuki and Yasumoto, 2000, Vale et al., 2002) for both the identification and quantitative determination of algal toxins.

Although OA and its methylated analogues DTX-1 and dinophysistoxin-2 (DTX-2) have been identified as being responsible for most DSP outbreaks (Carmody et al., 1996, Lawrence and Scoot, 1993, Luckas, 1992, Vale and Sampayo, 1999, Van Egmond et al., 1993), much less attention has been paid to the acyl derivatives of DSP toxins, also referred to as dinophysistoxin-3 (DTX-3). Due to their high molecular weight and lipophilicity, they cannot be directly detected with adequate sensitivity both by the fluorometric procedure of Lee et al. (1989) and by LC-MS/MS, but an alkaline hydrolysis reaction to release fatty acids from the parent toxins may contribute to a better estimation of their abundance. On the other hand, the lack of standards for these substances poses a further problem in their analytical determination (Vale and Sampayo, 2002).

Given the above, there is an apparent need for development of detoxification methodologies of shellfish from DSP toxins. Such methods should be rapid, efficient, and easy to apply and should not alter the quality and sensory properties of shellfish. Methods investigated so far for this purpose include thermal processing, freezing, evisceration, supercritical CO_2 with acetic acid, irradiation and ozonation (Gonzalez et al., 2002; Reboreda et al., 2010).

Ozone has been primarily applied for the treatment of drinking water (Bryant et al., 1992) as well as municipal and industrial waste water (Stover and Jarnis, 1981). Preliminary studies by Gacutan et al. (1984, 1985) on the use of ozone as a means of shellfish detoxification demonstrated that ozone gas may effectively inactivate PSP toxins from *Perna viridis* contaminated by *Pyrodinium bahamense*. However, results of subsequent study by White et al. (1985) were totally contradictory to previous studies in the context that no detoxification occurred in *Mya arenaria* exposed to ozone treatment. Schneider and Rodrick (1995) studied potential reduction of toxins associated with Florida's red tide organism, *Gymnodinium breve*. Results showed a three log reduction in the total amount of toxin recovered after 10 min (135 ppm) of exposure to ozone. Rositano et al. (1998) studied destruction of cyanobacterial toxins by ozone and concluded that free peptide hepatotoxins were rapidly destroyed by ozone in alkaline pH environments.

Despite the fact that ozone treatment has so far been, more or less, effectively applied for decontamination of shellfish from several marine biotoxins, a rather limited number of studies has dealt with lipophilic toxins and more precisely DSP toxins (Croci et al., 1994, Reboreda et al., 2010). These applications, however, involved treatment of the seawater where shellfish grow, rather than the end product itself, so may only apply before shellfish harvesting. Ozone has been declared by the USFDA in 2001 as a GRAS food preservative, approved for direct contact with foods (Khadre et al. 2001).

Based on the above, the first objective of the present study was the correlation of results of the three methods, (a) HPLC method, using ADAM as the fluorescent derivatizing reagent, (b) a LC-MS/MS method and c) the reference MBA used to detect and/or quantify OA, DTX-1 and its acyl derivatives in mussels. All samples analyzed originated from the DSP toxic episodes of years 2006 and 2007 in Greece. The second objective was to investigate the potential reduction of DSP toxins using gas ozone in mussels, post-harvest.

MATERIALS AND METHODS

Samples

For the first part of the study, mussels (*Mytillus galloprovincialis*) were collected during the DSP episodes of 2006 and 2007 from several sampling stations belonging to three different production areas of Greece: the Gulf of Thermaikos in northern Greece (Thessaloniki, Pieria, Imathia), the Gulf of Maliakos in central Greece (Fthiotida), and the Gulf of Saronikos in southern Greece (Megara) (Figure 1). A small number of market samples were also tested at the same time periods (Table 1). A total of 103 samples, originating from the Greek National Monitoring Program for Marine Biotoxins (National Reference Laboratory on Marine Biotoxins of Greece), were tested upon arrival to the laboratory by the MBA method. Selected samples, both positive and negative according to the MBA were stored at -70°C until chemical analyses. The hepatopancreas (HP) of mussels was used in all methods employed, as the lipophilic DSP toxins mainly accumulate in this organ (Murata et al., 1982). All samples were analyzed in duplicate.

Figure 1. Map of Greece showing specific mussel sampling locations (Gulf of Thermaikos, Gulf of Maliakos and Gulf of Saronikos).

For the second part of the study, 21 mussel samples, derived from the Greek National Monitoring Program for Marine Biotoxins (National Reference Laboratory on Marine Biotoxins of Greece, NRLMB) and originating from sampling stations mentioned above, were tested upon arrival to the laboratory by the MBA method, using the HP tissue. Selected samples, both positive and negative according to the MBA, were stored at –70°C until chemical analyses by LC-MS/MS in order to characterize toxin content and profile before further processing (ozonation), in the form of either whole shucked mussels or homogenized whole tissue. After ozanation, remaining toxicity of all samples was tested by MBA, whereas LC-MS/MS analysis was also carried out in order to quantify the effect of each treatment on toxin content and profile.

Reagents

Solvents used for the MBA were Tween-60 grade "for synthesis" (Sigma, Sigma-Aldrich, St. Louis, MO, USA) and acetone (Merck, Darmstadt, Germany), analytical grade.

Solvents used for the HPLC-FLD and LC-MS/MS analyses were LC grade methanol, chloroform, acetone, water, acetonitrile, and n-hexane (Merck, Darmstadt, Germany). The OA standard solution (NRC CRM-OA-b, Institute for Marine Biosciences, Canada) was used for the preparation of the calibration curve for both in HPLC-FLD and LC-MS/MS methodology. A certified Reference Material Mussel Tissue with a certified value of 10.1 μg OA/g and 1.3 μg DTX-1/g (NRC CRM-DSP-Mus-b, Institute for Marine Biosciences, National Research Council of Canada, and Halifax, Canada) was used for recovery determination both in HPLC-FLD and LC-MS/MS methods. The ADAM from Serva (Heidelberg, Germany) was used for derivatization of both standard solutions and samples. All standards, reference materials and the derivatization reagent were stored at –20°C. Solid Phase Extraction (SPE) cartridges packed with silica were purchased from Alltech (Deerfield, USA).

Mouse Bioassay

The MBA for the determination of DSP toxicity in samples was performed as described by Yasumoto et al. (1978). Briefly, 20 g portion of HP was extracted three times with 50 ml acetone each time and filtered through a cellulose filter. The combined toxin extract was evaporated to dryness and resuspended in 4 ml of 1% Tween-60 (5 g HP/ml). Each one of three mice (Albino Swiss, 18–20 g body weight) was injected intraperitoneally with 1 ml of this solution.

Samples toxicity after ozone gas treatment was determined using the EU-harmonized protocol [Version 4.0, (33)]. An aliquot of 100 g of whole flesh tissue homogenate was weighed into 500 ml plastic centrifuge containers (Sigma Laborzentrifugen GmbH, Osterode am Harz, Germany), to which acetone (300 ml) was added. The mixture was homogenized for 2 min using an Ultra turrax (15 mm shaft, 10,000 rpm; IKA, Germany) and centrifuged at 1200 × g for 10 min at 4°C (Centrifuge Sigma 4K15C). The supernatant was transferred through a filter paper into a labeled 1 l round-bottom flask, while the tissue residue was re-extracted in the same way with 200 ml of acetone. The combined 500 ml filtrate was evaporated under vacuum at 42 ± 2°C

(Rotavapor R-200, Buchi, Switzerland), until complete removal of acetone. The aqueous residue of the round-bottom flask was transferred to a 100 ml cylinder and volume was adjusted to 100 ml by addition of de-ionized water, while 100 ml of diethyl ether were used to rinse off the residue of the round-bottom flask. Both aqueous and diethyl ether extracts were transferred to a 500 ml separatory funnel. The mixture was agitated and following separation of the aqueous and diethyl ether phases, the aqueous layer was transferred back into the round-bottom flask. The water was re-extracted twice with 100 ml of diethyl ether and separated in the separatory funnel yielding ca. 300 ml of diethyl ether extract, which was backwashed twice with 20 ml de-ionized water. The ether phase was collected in another round-bottom flask and rotary evaporated to dryness under vacuum at 42 ± 2°C. The dry residue was resuspended in 1% Tween-60 to a final volume of 4 ml (25 g whole flesh/ml). When necessary, the resuspended extract was further homogenized using an Ultra turrax (8 mm shaft) prior to injection. Each one of three mice (Albino Swiss, 18-20 g body weight) was injected intraperitoneally with 1 ml of this solution.

In both protocols, the criterion of toxicity established by the EU Regulation 2074/2005/EC was employed, which is the death of two out of three mice within 24 hr of inoculation with an extract equivalent to 5 g of HP. This constitutes a positive result for the presence of the lipophilic toxins mentioned in the above Regulation, including OA, DTX-1, -2 and -3. The mice were allowed laboratory feed and water *ad libitum* throughout the observation period. All animal manipulations were performed in accordance with the EU Directive 86/609/EC (1986), under official license from the Prefectural Veterinary Service of Thessaloniki, Greece.

Toxin Extraction for HPLC-FLD and LC-MS/MS

Toxin extraction was performed according to the method of Mouratidou et al. (2004). Briefly, 1 g of homogenized digestive glands (DG) (hepatopancreas) was vortex-extracted with 4 ml of 80% aqueous methanol for 1 min, and centrifuged at 4000 rpm for 5 min. A 2.5 ml supernatant aliquot was rinsed twice with 2 ml of hexane, then 0.5 ml of water was added and further extracted twice with 2 ml chloroform. The aqueous phase was discarded whereas the combined chloroform phase volume was corrected to 10 ml in 10 ml volumetric flask.

Derivatization and Clean-up Procedure for HPLC-FLD

For OA derivatization an aliquot (0.5 ml) of either sample extract (in chloroform) or calibration standard, was transferred into 25 ml amber plastic vials and dried under N_2. The residues were esterified with 200 μl of 0.2% ADAM solution for 1 hr, in the dark, at 35°C. The ADAM solution was prepared daily, by dissolving ADAM (5 mg) in acetone (100 μl) and volume was adjusted to 2.5 ml with methanol. After evaporating the solvent, the reaction products were redissolved in 2 ml hexane-chloroform mixture (1:1) in three portions, and loaded to a SPE silica cartridge (650 mg, Alltech), preconditioned with 6 ml chloroform followed by 3 ml hexane-chloroform mixture (1:1). The sample passed slowly (1 drop/sec) through the clean up column, which was then washed with 5 ml of the same solvent followed by 5 ml of chloroform. The ADAM-OA esters were then eluted with 5 ml of chloroform-methanol mixture (95:5) and evaporated to dryness under N_2. The residue was reconstituted in methanol (0.2 ml) for HPLC-FLD analysis (Mouratidou et al., 2004).

HPLC-FLD Analysis

The HPLC-FLD analyzes were performed on Shimadzu HPLC system, model LC-10AD (Columbia, USA), equipped with wavelength fluorescence detector (model RF-10A XL) set at 365 nm excitation and 415 nm emission wavelength. Identity of ADAM-OA and ADAM-DTX-1 peaks was confirmed by matching the retention time (RT) with standard solutions of OA and DTX-1, respectively as well as standard solution (CRM). Separation of the ADAM-OA derivative was achieved on 250-4Lichrospher 100 RP-18 (5μm) column, (Merck), at 35°C. Separation of the ADAM-OA derivative was achieved on 250-4Lichrispher 100 RP-18 (5μm) column, (Merck), at 35°C, using an isocratic solvent mixture of acetonitrile: water (80:20) for 15 min. The flow rate was 1ml/min and the injection volume was 20μl. The derived calibration curve (5 points obtained from the average of duplicate injections and triplicate derivatizations for each concentration level) showed good linearity in the concentration range of 1–10 ng OA on-column. The OA calibration curves were also used for the calibration of DTX-1 measurements, since under the isocratic conditions applied, the relative molar responses for the two toxins are identical considering that the fluorescence signal is derived only from the anthracenyl moiety (Mouratidou et al., 2004).

Sample Preparation for Determination of DSP Toxins by LC-MS/MS

Additional sample preparation for LC-MS/MS analyses was based on the method of Vale et al. (2002), with some modifications. For determination of the free forms of DSP toxins, after toxin extraction (as described under section 2.4), 4 ml of the chloroform phase were removed and evaporated to dryness under N_2. The residue was reconstituted in methanol (0.5 ml) for LC-MS/MS analysis. For determination of total DSP toxins (free plus esterified forms), a hydrolysis procedure was conducted according to the method of Vale et al. (2002). In one milliliter aliquots of the 80% methanol extracts (equivalent to 0.2 g HP, see section 2.4), 400 μl of 1 M NaOH in methanol: water (9:1) at 35°C for 40 min were added and incubated. The mixture was neutralized with 415 μl of 1 M HCl. The neutralized solution was rinsed twice with 1 ml of hexane, then 0.5 ml of water was added and further extracted twice with 1 ml chloroform. This chloroformic solution was evaporated to dryness under N_2 and the residue was reconstituted in methanol (0.5 ml) for LC-MS/MS analysis). Aliquots of 20 μl were injected into the LC-MS/MS for both free and total DSP toxins analysis.

LC-MS/MS Analysis

Liquid chromatography-mass spectrometry was performed on an Agilent LC 1100 series liquid chromatograph coupled to an Agilent LC/MSD Trap SL (two octapoles) (Stuttgart, Germany) equipped with an atmospheric pressure electrospray ionization (ESI) interface. Separation was achieved on an Eclipse XDB-C18 column (4.6 mm × 150 mm, 5μm) at 30°C. Mobile phase A was 95% aqueous ACN and mobile phase B was 100% water, both containing 2 mM ammonium formate and 50 mM formic acid (Fux et al., 2007). Gradient elution from 40 to 90% A was performed over 3 min, held at 90% A for 7 min and returned to 40% A in 2 min. The flow rate was 0.6 ml/min and the injection volume was 20 μl. The electrospray capillary was set at 4 kV, the nebulizer at 50 psi, dry gas at 10 l/min, and dry temperature at 350°C. Toxin detection and quantification was carried out using selected ion monitoring (SIM) of negatively charged ions for [M-H] of OA (*m/z* 803.5) and DTX-1 (*m/z* 817.5). Detection was based on daughter ions monitored *m/z* namely: 254.9, 563.1, 785.2 for OA and *m/z* namely: 254.9, 563.1, and 799.2 for DTX-1.

Ozone Treatment

Ozone was generated by exposure of pure oxygen gas supplied by a 10m³ cylinder to high voltage electrical discharge. The ozone generator C-Lasky series (C-L010-DTI/C-L010-DSI, AirTree Ozone Technology Co., Baunatal, Germany) was used for the generation of ozone. Mussel samples were ozonated inside a refrigerator (4°C) at a dose of gas ozone of 15 ppm and exposure time of 6 hr. Ozone gas was fed into the refrigerator by means of plastic (PVC) tubing, whereas during treatment, samples (either whole shucked or homogenized) were placed in glass trays in order to achieve the largest possible exposure of the samples to ozone gas. Ozone concentration inside the refrigerator was measured using a model OS-4 ECO-Sensors ozone meter (Santa Fe, USA).

DISCUSSION AND RESULTS

Mouse Bioassay

The results from MBA analysis are shown in Table 1. A total of 89 samples were found positive whereas 14 samples were negative to DSP toxins. This is one of the very few studies conducted in Greece for the detection and/or quantification of OA, DTX-1 (Mouratidou et al., 2004 and 2006, Reizopoulou et al., 2008) and their esters (Prassopoulou et al., 2009) and the first to include samples collected from sampling stations in the Gulf of Maliakos.

HPLC-FLD

The mean RT for OA was 9.8 ± 0.1 min and for DTX-1 11.8 ± 0.1 min. Typical chromatographic profiles of certified reference material and contaminated sample from the area of Thessaloniki are illustrated in Figure 2(a) and 2(b) respectively.

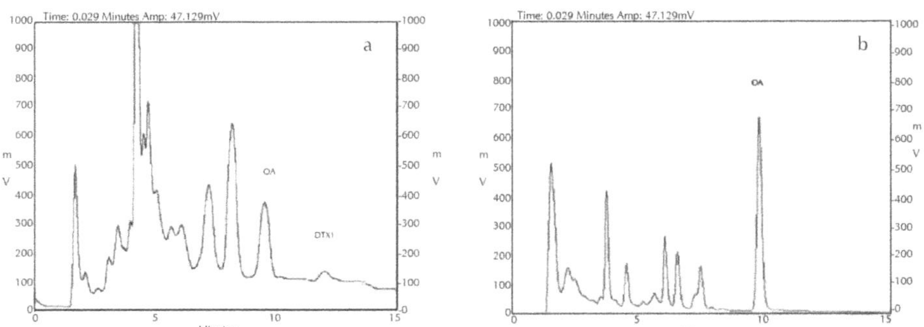

Figure 2. Chromatographic profile of toxins by HPLC-FLD (a) standard mussel containing OA and DTX-1, (b) sample from Thessaloniki containing only OA.

A calibration curve for OA was constructed in the range of 1–10 ng OA on column. The calibration curve showed excellent linearity $R^2 = 0.9928$. To determine the recovery of the HPLC analytical procedure using the ADAM fluorescent reagent, a Certified Reference Material (NRC CRM-DSP-Mus-b) containing 10.1 ± 0.8 μg OA/g and 1.3 ± 0.2 μg DTX-1/g was used. The CRM homogenate, diluted four times at the extraction stage in order to fit within the calibration curve range, was repeatedly analyzed (n = 3). The average peak area of OA, after multiplication by the dilution factor, corresponded to 9.8 μg/g which is equal to 97% of the target value. The average peak area of DTX-1 (quantified using the OA calibration curve and assuming that the relative molar responses for both toxins were equal), after multiplication by the dilution factor, corresponded to 1.2 μg/g which is equal to the 92% of the target value. The recovery of the method is within the range of 95 ± 5% reported by Kelly et al. (1996) and 91-99% reported by Gonzalez et al. (2000). The LOD and LOQ for OA was calculated according to the IUPAC criterion (signal-to-noise ratio of 3) and was 0.015 μg/g HP and 0.050 μg/g respectively.

The results from HPLC-FLD analysis are shown in Table 1. The concentrations of free OA ranged from 0.40 to 62.67 μg/g HP (88 samples). Also, there were samples in which free OA was not detected or OA concentrations were below the detection limit of the method (0.015 μg/g HP). The DTX-1 was not detected in any of the samples.

Mouratidou et al, (2004) determined OA and DTX-1 in mussels harvested from the Gulf of Thermaikos during the 2002 DSP episode in Greece using HPLC with fluorometric detection (FLD). In almost all samples tested OA was determined at levels remarkably high (max. value 36.06 μg/g HP) exceeding by far the regulatory limit of 0.80 μg/g HP. The DTX-1 was found only in one sample at a concentration of 0.51 μg/g HP. Levels of OA in the above study were generally lower than those determined in the present study.

Prassopoulou et al. (2009) determined OA and its polar and non-polar esters in mussels harvested from the Gulfs of Thermaikos and Saronikos during the 2006–2007 DSP episode in Greece using HPLC-FLD after hydrolysis of the samples with NaOH. In the 50 samples tested free OA ranged from 0.122 to 4.099 μg/g shellfish meat, corresponding to 0.61 and 20.50 μg/g HP, with 98% of samples exceeding the regulatory limit for OA. These values are generally lower than those recorded in the present study ranging between 0.40 and 62.67 μg/g HP.

Table 1. Concentration of okadaic and acyl derivatives determined by the applied methods (MBA, HPLC-FLD, LC-MS/MS) of mussels' hepatopancreas.

						LC-MS/MS			
Sample	**Sampling Region**	**Sampling Date**	**Sampling Year**	**Mouse Bioassay (mice deaths)**	**HPLC-FLD μg OA/g**	**μg OA/g**	**μg esters OA/g**	**% esters over total OA**	**Total OA (μg/g)**
1	Thessaloniki	10/1	2007	Positive (+++)	16.00	16.40	5.60	25.45	22.00
2	Thessaloniki	10/1	2007	Positive (+++)	44.00	44.00	ND	0.00	44.00
3	Thessaloniki	10/1	2007	Positive (+++)	56.80	57.60	ND	0.00	57.60

Table 1. *(Continued)*

Sample	Sampling Region	Sampling Date	Sampling Year	Mouse Bioassay (mice deaths)	HPLC-FLD μg OA/g	LC-MS/MS μg OA/g	μg esters OA/g	% esters over total OA	Total OA (μg/g)
4	Thessaloniki	22/1	2007	Positive (+++)	41.13	41.00	24.40	16.55	65.40
5	Thessaloniki	25/1	2007	Positive (+-+)	30.00	30.27	19.60	17.75	49.87
6	Thessaloniki	25/1	2007	Positive (+++)	24.70	24.53	11.60	13.62	36.13
7	Thessaloniki	1/2	2007	Positive (++-)	36.87	36.67	11.40	9.39	48.07
8	Thessaloniki	9/2	2007	Positive (+++)	23.80	24.20	ND	0.00	24.20
9	Thessaloniki	13/2	2007	Positive (++-)	18.20	18.20	ND	0.00	18.20
10	Thessaloniki	15/3	2007	Positive (+++)	31.53	31.87	13.60	12.45	45.47
11	Thessaloniki	20/3	2007	Positive (-++)	3.40	3.20	3.60	52.94	6.80
12	Thessaloniki	27/3	2007	Positive (+++)	20.00	19.60	3.80	16.24	23.40
13	Thessaloniki	27/3	2007	Positive (+++)	8.60	9.00	ND	0.00	9.00
14	Thessaloniki	10/4	2007	Positive (++-)	14.80	14.40	0.60	3.00	20.00
15	Thessaloniki	11/4	2007	Positive (-++)	11.60	11.00	ND	0.00	11.00
16	Thessaloniki	12/4	2007	Positive (+++)	27.60	27.00	12.6	31.82	39.60
17	Thessaloniki	17/4	2007	Positive (-++)	24.00	25.00	5.80	18.83	30.80
18	Thessaloniki	17/4	2007	Positive (+++)	5.20	5.40	5.00	48.08	10.40
19	Thessaloniki	17/4	2007	Positive (+-+)	14.00	14.20	5.80	29.00	20.00
20	Thessaloniki	17/4	2007	Positive (+++)	12.00	11.00	ND	0.00	11.00
21	Thessaloniki	21/8	2007	Positive (+++)	27.20	27.60	ND	0.00	27.60
22	Thessaloniki	27/8	2007	Positive (++-)	29.20	29.26	ND	0.00	29.26
23	Thessaloniki	18/9	2007	Positive (+++)	6.00	6.20	ND	0.00	6.20
24	Thessaloniki	25/9	2007	Positive (+++)	62.67	62.07	ND	0.00	62.07
25	Thessaloniki	2/10	2007	Positive (+++)	13.00	13.40	ND	0.00	13.40
26	Pieria	3/1	2007	Positive (+-+)	9.80	10.20	3.00	22.73	13.20
27	Pieria	3/1	2007	Positive (+++)	35.20	35.00	9.20	20.81	44.20
28	Pieria	15/1	2007	Positive (+++)	26.00	25.20	ND	0.00	25.20
29	Pieria	15/1	2007	Positive (+++)	15.60	15.20	26.60	63.64	41.80
30	Pieria	15/1	2007	Positive (+++)	26.20	25.80	17.40	40.28	43.20
31	Pieria	15/1	2007	Positive (+++)	51.33	50.07	17.80	10.60	67.87
32	Pieria	15/1	2007	Positive (+++)	48.67	48.87	9.40	6.03	58.27
33	Pieria	15/1	2007	Negative	ND	ND	ND	N/A	ND
34	Pieria	24/1	2007	Positive (+++)	54.00	53.87	11.40	6.59	65.27
35	Pieria	24/1	2007	Positive (+++)	50.00	49.20	12.40	7.75	61.60
36	Pieria	24/1	2007	Positive(+++)	57.33	57.47	12.80	6.91	70.27
37	Pieria	24/1	2007	Positive (+++)	12.20	11.60	3.80	24.68	15.40
38	Pieria	24/1	2007	Positive (+++)	77.60	78.20	9.60	10.93	87.80
39	Pieria	29/1	2007	Positive (+++)	33.33	34.00	11.80	10.37	35.80
40	Pieria	29/1	2007	Positive(+++)	24.20	23.20	4.20	15.33	27.40
41	Pieria	29/1	2007	Positive (+-+)	13.20	14.00	2.00	12.50	16.00
42	Pieria	29/1	2007	Positive (+++)	20.60	20.40	8.80	30.14	29.20
43	Pieria	29/1	2007	Negative	ND	ND	ND	N/A	ND
44	Pieria	5/2	2007	Negative	ND	ND	ND	N/A	ND

Table 1. *(Continued)*

Sample	Sampling Region	Sampling Date	Sampling Year	Mouse Bioassay (mice deaths)	HPLC-FLD µg OA/g	LC-MS/MS µg OA/g	µg esters OA/g	% esters over total OA	Total OA (µg/g)
45	Pieria	5/2	2007	Positive (+++)	26.00	24.40	2.60	9.63	27.00
46	Pieria	5/2	2007	Positive (+++)	8.00	8.40	0.20	2.33	8.60
47	Pieria	5/2	2007	Positive (-++)	6.60	7.00	3.00	30.00	10.00
48	Pieria	20/2	2007	Positive (-++)	3.80	4.40	ND	0.00	4.40
49	Pieria	19/3	2007	Positive (-++)	10.80	11.20	5.80	34.12	17.00
50	Pieria	26/3	2007	Positive (+++)	10.80	10.60	ND	0.00	10.60
51	Pieria	2/4	2007	Positive (+++)	10.20	10.20	14.20	58.20	24.40
52	Pieria	2/4	2007	Positive(+++)	8.40	8.40	1.20	12.50	9.60
53	Pieria	17/4	2007	Positive (+-+)	3.80	4.20	2.40	34.29	7.00
54	Pieria	17/4	2007	Positive (+++)	10.80	10.60	ND	0.00	10.60
55	Imathia	5/3	2007	Negative (+--)	ND	ND	ND	N/A	ND
56	Imathia	12/3	2007	Positive (+++)	9.00	8.80	3.80	30.16	12.60
57	Imathia	12/3	2007	Positive (+++)	50.00	49.20	8.00	13.99	57.20
58	Imathia	12/3	2007	Positive (+++)	28.00	27.00	ND	0.00	27.00
59	Imathia	19/3	2007	Positive (+++)	8.00	7.20	ND	0.00	7.20
60	Imathia	19/3	2007	Positive (+++)	7.00	6.60	ND	0.00	6.60
61	Imathia	19/3	2007	Positive (+++)	40.00	38.80	22.20	36.39	61.00
62	Imathia	27/3	2007	Positive (+++)	52.00	50.20	17.80	26.18	68.00
63	Imathia	27/3	2007	Positive (+++)	25.60	25.00	1.60	6.02	26.60
64	Imathia	27/3	2007	Positive (+++)	17.00	16.60	9.60	36.64	26.20
65	Imathia	2/4	2007	Positive (+++)	25.80	26.40	12.60	32.31	39.00
66	Imathia	2/4	2007	Positive (+++)	17.20	17.80	6.00	25.21	23.80
67	Imathia	2/4	2007	Positive(+++)	17.60	17.80	1.60	8.25	19.40
68	Imathia	11/4	2007	Positive (+++)	38.00	36.60	16.00	30.42	52.60
69	Imathia	11/4	2007	Positive (+++)	6.60	5.80	ND	0.00	5.80
70	Imathia	11/4	2007	Positive (+++)	12.00	11.40	4.00	25.97	15.40
71	Imathia	17/4	2007	Positive (+++)	6.00	6.20	ND	0.00	6.20
72	Imathia	17/4	2007	Positive (+++)	13.40	13.00	7.80	37.50	20.80
73	Imathia	17/4	2007	Positive (+++)	37.20	34.00	3.40	9.09	37.40
74	Imathia	26/4	2007	Positive (+++)	4.20	4.00	3.20	44.44	7.20
75	Imathia	14/5	2007	Positive (+-+)	1.20	1.60	ND	0.00	1.60
76	Fthiotida	24/7	2007	Positive (++-)	12.80	12.40	4.20	25.30	16.60
77	Fthiotida	1/8	2007	Positive (+++)	22.00	21.60	0.80	3.57	22.40
78	Fthiotida	14/8	2007	Positive (+++)	12.40	13.00	ND	0.00	13.00
79	Megara	16/1	2007	Positive (+++)	14.00	16.00	27.80	63.47	43.80
80	Megara	13/2	2007	Positive (+++)	46.60	46.40	34.40	48.59	70.80
81	Megara	27/2	2007	Positive (+++)	53.80	52.20	1.40	2.61	53.60
82	Market	17/4	2007	Positive (+++)	9.20	8.80	3.40	27.87	12.20
83	Market	9/3	2007	Negative (--+)	ND	ND	ND	N/A	ND
84	Market	9/3	2007	Positive (+++)	10.00	9.60	5.80	37.66	15.40
85	Market	13/3	2007	Positive (+++)	7.60	7.20	3.40	32.08	10.60

Table 1. *(Continued)*

Sample	Sampling Region	Sampling Date	Sampling Year	Mouse Bioassay (mice deaths)	HPLC-FLD µg OA/g	LC-MS/MS µg OA/g	µg esters OA/g	% esters over total OA	Total OA (µg/g)
86	Market	13/3	2007	Negative (+--)	ND	ND	ND	N/A	ND
87	Market	13/3	2007	Positive (+-+)	12.80	12.40	5.80	31.87	18.20
88	Fthiotida	13/3	2006	Negative	ND	ND	ND	N/A	ND
89	Fthiotida	1/6	2006	Negative	0.40	0.60	ND	0.00	0.60
90	Fthiotida	2/6	2006	Negative (-+-)	0.50	0.60	ND	0.00	0.60
91	Fthiotida	2/6	2006	Positive (-++)	ND	ND	ND	N/A	ND
92	Fthiotida	2/6	2006	Negative	ND	ND	ND	N/A	ND
93	Fthiotida	6/6	2006	Positive (+-+)	0.90	1.00	ND	0.00	1.00
94	Fthiotida	6/6	2006	Negative	ND	ND	ND	N/A	ND
95	Fthiotida	27/6	2006	Negative	ND	ND	ND	N/A	ND
96	Fthiotida	27/6	2006	Positive (++-)	6.60	7.00	ND	0.00	7.00
97	Fthiotida	27/6	2006	Positive (++-)	ND	ND	ND	N/A	ND
98	Fthiotida	13/9	2006	Negative (-+-)	ND	ND	ND	N/A	ND
99	Fthiotida	19/9	2006	Negative (-+-)	ND	ND	ND	N/A	ND
100	Fthiotida	21/9	2006	Positive (++-)	ND	ND	ND	N/A	ND
101	Megara	31/10	2006	Positive (+++)	23.60	23.40	2.80	10.69	26.20
102	Megara	14/12	2006	Positive (+++)	28.60	28.60	2.80	8.92	31.40
103	Megara	28/12	2006	Positive (+++)	24.40	24.00	ND	0.00	24.00

ND: not detected

LC-MS/MS

The mean RT for OA was 8.0 ± 0.1 min and 10.1 ± 0.5 for DTX-1 in the LC-MS/MS analysis. Typical chromatographic profile of certified reference material and MS/MS spectra of OA and DTX-1 are illustrated in Figure 3.

Toxins were detected using SIM of negatively charged ions for $[M-H]^-$ of OA (*m/z* 803.5) and DTX-1 (*m/z* 817.5). Detection was based on daughter ions monitored *m/z* namely: 254.9, 563.1, and 785.2 for OA and *m/z* namely: 254.9, 563.1, and 799.2 for DTX-1.

Calibration curves for the LC-MS/MS analysis were in the range of 2-10 ng OA on column. The calibration curve showed excellent linearity $R^2 = 0.9959$. The CRM homogenate, diluted four times at the extraction stage in order to fit within the calibration curve range, was repeatedly analyzed (n = 3). The average peak area of OA after multiplication by the dilution factor corresponded to 9.7 µg/g, which is equal to 96% of the target value. The average peak area of DTX-1 (quantified using the OA calibration curve and assuming that the relative molar responses for both toxins were equal), after multiplication by the dilution factor, corresponded to 1.2 µg/g which is equal to the 92% of the target value for LC-MS/MS. The recovery of the method is near the values of 98-99% reported by Goto et al. (2001) and ca. 100% Suzuki and Yasumoto (2000). The LOD and LOQ (according to the IUPAC criterion) were 0.045

μg/g HP and 0.135 μg/g HP respectively. Goto et al. (2001) determined marine toxins associated with diarrhetic shellfish by LC-MS. They showed that the method quantified toxins at low concentrations and was suitable as a warning for shellfish toxicity. The detection limits in the muscle and DG were as follows: 5 and 10 ng/g for OA, 10 and 20 ng/g for DTX-1 and 20 and 40 ng/g for OA and DTX-1 esters.

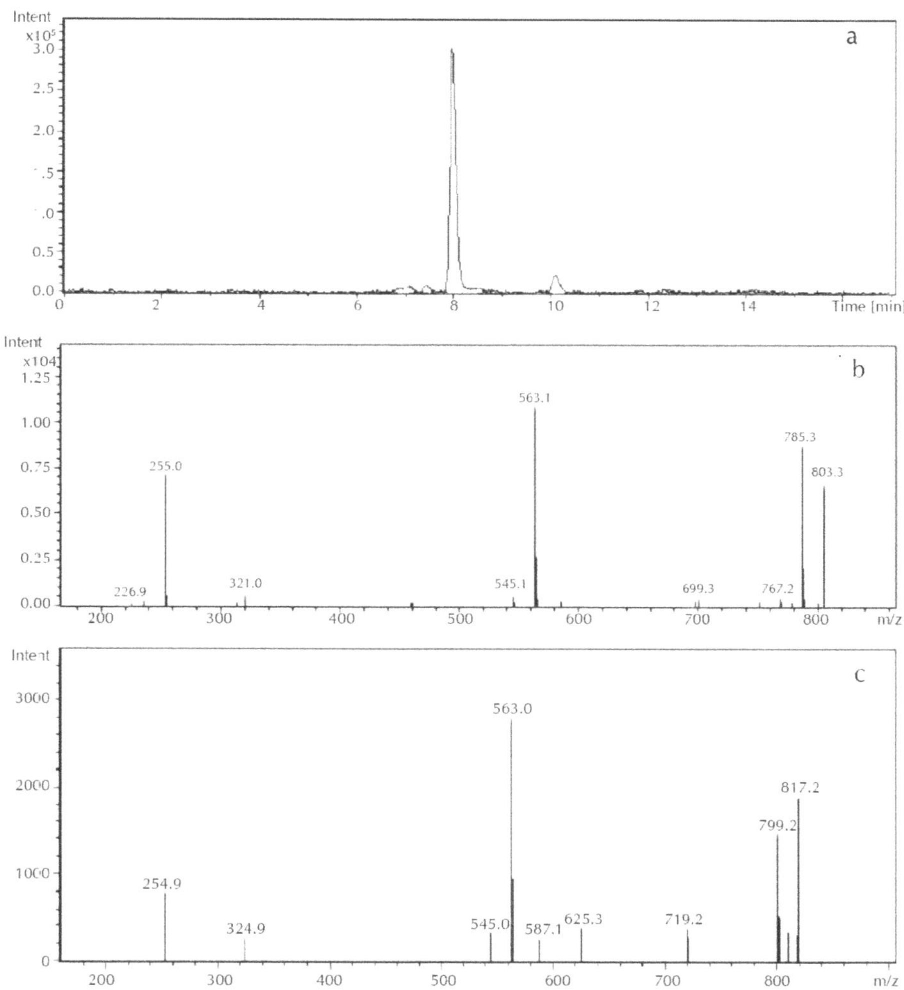

Figure 3. Chromatographic profile of toxins by LC-MS/MS (a) standard mussel containing OA and DTX-1, (b) negative electrospray MS/MS spectrum of OA, and (c) negative electrospray MS/MS spectrum of DTX-1.

The results of LC-MS/MS analysis are shown in Table 1. The concentrations of free OA ranged from 0.56 to 62.07 μg/g HP. Also, there were samples in which OA was not detected or OA concentrations were under the detection limit of the method (0.045 μg/g HP). The DTX-1 was not detected in any of the samples. The concentration of OA esters ranged from 0.20 to 34.4 μg/g HP and the total OA concentration (free OA plus OA esters) ranged from 1.00 to 87.80 μg/g HP. Within the European Union, the maximum permissible level of OA, DTX's and pectenotoxins is laid down to 160 μg of OA equivalents/kg mussel tissue (or 800 μg of OA equivalents/kg HP) (Regulation 853/2004/EC).

The OA esters determined by Prassopoulou et al. (2009) in Greek mussels by HPLC-FLD ranged between ND (not detectable) and 44903.5 μg/kg of shellfish meat corresponding to 224.5 μg/g HP respectively. These esters' values are much higher than those of the present study ranging between non detectable and 34.40 μg/g HP as determined by LC-MS/MS. To the best of our knowledge there are no other reports on esterified forms of DSP toxins in Greece. The large majority of esters detected in the above study belonged to the polar group where as some non-polar esters were also found. The DTX-1 was not found in any of the samples tested. The absence of DTX-1 in the above samples as in those of the present study may be attributed to differences in dominant *Dinophysis* species occurring between different regions. *Dinophysis acuta* was the dominant species causing the 1995 and 1998 DSP outbreaks in Portugal, where more non-polar esters of OA were quantified (Vale and Sampayo, 1999) while *Dinophysis acuminata* was identified as the main *Dinophysis* species present in the Gulf of Thermaikos (Koukaras and Nikolaidis, 2004).

Vale and Sampayo (2002) determined OA and DTX-2 in Portuguese shellfish (mussels and cockles in 2000) using LC-MS (single quadrupole MS-ESI interface operated in the negative ion mode). The methodology used had a LOD 80–200-fold lower than maximum regulatory levels set. The OA was determined to be 0.026 μg/g edible parts (EP) before hydrolysis and 0.061 μg/g EP after hydrolysis with NaOH for mussels. Respective values for cockles were non detectable and 0.085 μg/g DTX-1 was not detected in any of the samples tested. Oysters, clams, carpet shells, and razor clams presented a smaller health risk for consumers.

Chapela et al. (2008) determined OA, DTX-1, DTX-2, YTX, PTX-2, AZA-1, and SPX-1 in seven different species of shellfish in fresh, frozen boiled and canned products using LC-MS/MS. Samples originated from Europe, and mostly from Spain. OA was the most frequently found toxin, appearing in mussels, cockles, clams, and scallops, followed by DTX-2 (in mussels, cockles and small scallops), yessotoxins (YTX) (in mussels) and finally azaspiracids (AZA-1) appearing only in small scallops. In a total of 12 samples OA, DTXs, and YTXs were quantified, although at levels under the legal limits allowed in the EU.

Draisci et al. (1995) determined DSP toxins in samples from north and south Adriatic Sea (Italy) using ionspray liquid chromatography-mass spectrometry (LC-ISP-MS). Concentrations of OA ranged from 0.32 to 1.51 μg/g HP while DTX-1 was found only in four samples (two samples from north and two samples from south Adriatic Sea) in concentrations ranging between 0.21 and 0.32 μg/g HP

Blanco et al. (2007) determined OA and DTX-2 and their conjugated forms in mussels from mussel farms located in Galician Rias (Spain) using HPLC-MS. This study determined mostly OA and only very small amounts of DTX-2. Average concentration of OA was 390 ng/g of digestive gland and 407 ng/g of DG for total OA (free plus conjugated forms).

Villar-Gonzalez et al. (2007) determined lipophilic toxins in samples from Spain from the 2005 toxic episode using LC-MS/MS. According to their study, 11 out of the 12 samples were contaminated with lipophilic toxins at levels above the European regulatory limit. The toxins found belonged mainly to the OA toxin group but no DTX-1 was found. A high percentage of esters in relation to the total amount of DSP toxins (35 to 79%) for mussels were also determined. Moreover, Villar-Gonzalez et al. (2008) determined OA and the acyl-derivative palOA and analogous compounds in Spanish shellfish. In the case of scallops, 89% of OA accumulated in the bivalve is acylated to generate palOA. However, only 27% of total OA-group toxin esters in razor clams were palmitoyl derivative.

Percentages of OA esters in Thermaikos Gulf ranged between 0 and 63.64% of the total OA detected, whereas the relevant percentages were 0-63.47% in Saronikos Gulf, 0-25.30% in Maliakos Gulf, and 27.87-37.66% for samples from market (Table 1). The respective percentages of OA esters reported by Vale and Sampayo (1999) and Prassopoulou et al. (2009) were generally higher than those of the present study and were in the range of 80-90%. The observed differences between different studies, as well as between different samples of the present study, could possibly arise from genetic differences in the mussels, as OA esterification is considered as an enzymatic mechanism involved in detoxification of DSP toxins (Vale and Sampayo, 1999, 2002).

Comparison of Three Methods

Although, in the HPLC-FLD method employed in the present study only free OA was determined, whereas total DSP toxins (free OA plus esters of OA) as well as other toxin groups are detected by the MBA, results of the two methods were comparable (Figure 4), with agreement reaching a level of 97.1%.

Additionally, comparison of the results obtained by HPLC-FLD and LC-MS/MS methods showed an excellent agreement (100%) with a regression analysis between the two methods resulting in $R^2 = 0.98$ (Figure 5). Keeping in mind that sample preparation for HPLC-FLD analysis included a multistep approach and given the instability of the ADAM derivatization agent, it is suggested that the LC-MS/MS analysis is preferable to the HPLC-FLD method for the determination of DSP toxins in mussels.

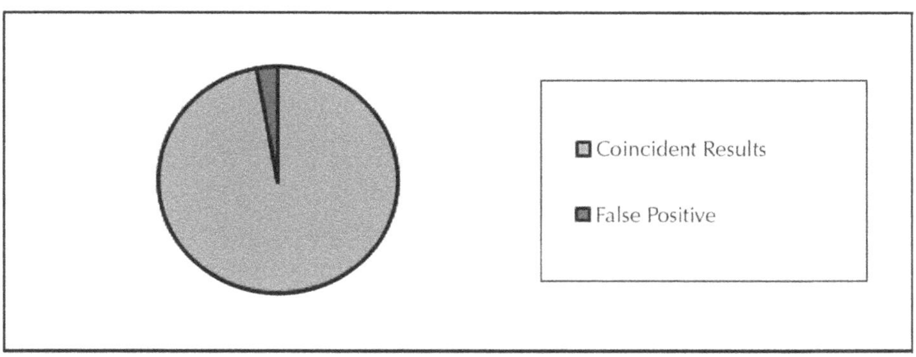

Figure 4. Comparison between mouse bioassay and high performance liquid chromatography method.

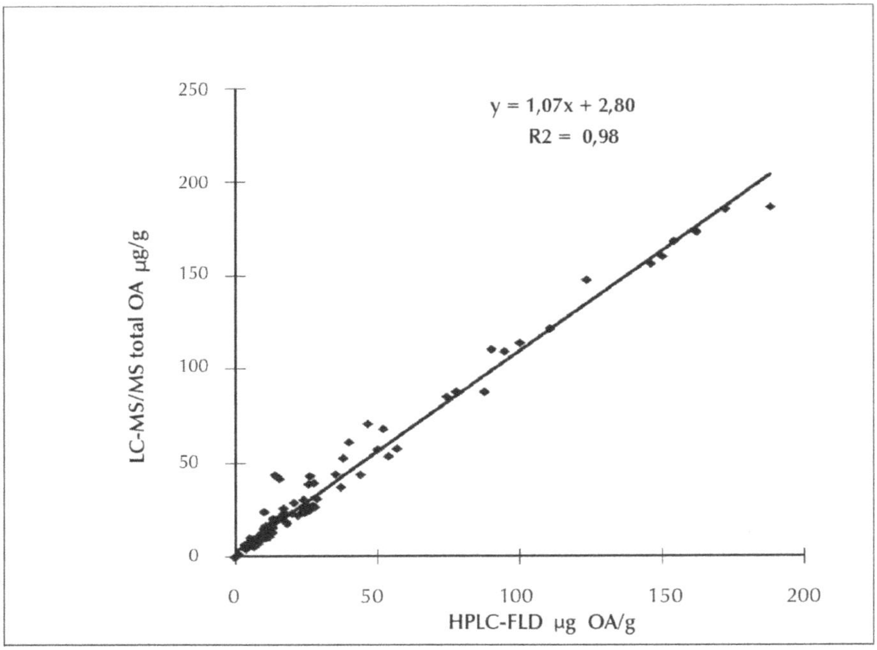

Figure 5. Correlation of LC-MS/MS vs. HPLC-FLD methods.

Comparison of results from the MBA and LC-MS/MS methods also showed a very good agreement (Figure 6), with the consistency of the two methods being 97.1%, as in the case of HPLC-FLD. There were only three samples in which OA was not detected (#91, 97, 98) by the LC-MS/MS, which were positive with the MBA This discrepancy could be explained by the fact that mouse death was due either to the presence of other lipophilic toxins in the samples for example YTX, AZA, GYM or SPXs or to the presence of free fatty acids in samples (false positives) (Suzuki et al., 2005). It should be noted that toxins such as GYM and SPX can cause the death of test animals in the bioassay, even at very low concentrations (Gerssen et al., 2008).

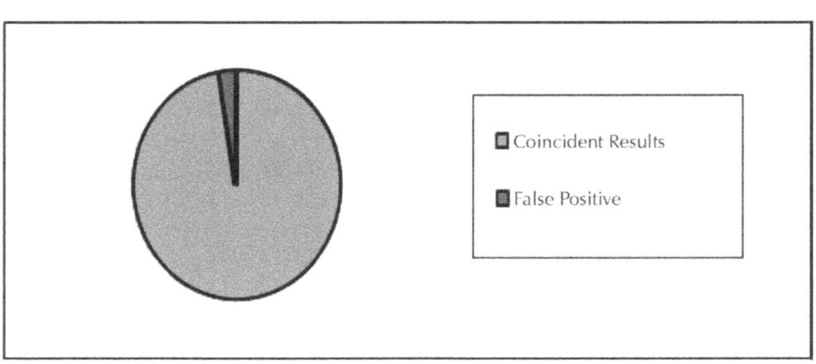

Figure 6. Comparison between MBA and LC-MS/MS method.

On the other hand, Turrell and Stobo (2008) compared the MBA with liquid chromatography-mass spectrometry for the detection of lipophilic toxins in shellfish from Scotland. In their study it was not possible to demonstrate complete equivalence of LC-MS with MBA, in contrast to the present study where MBA has shown almost complete agreement with LC-MS/MS.

Sample Ttreatment with Ozone

Results of decontamination trials by ozone gas treatment are presented in Table 2. Before treatment, toxin content of positive samples, for free OA, OA esters and total OA concentration (free OA plus OA esters) were in the range of 0.16–5.96 μg/g, ND to 6.84 μg/g and 0.98–12.80 μg/g mussel tissue, respectively, whereas in negative samples OA or OA esters were again non-detectable.

Despite the initial very high toxin content (OA and derivatives) of the samples included in the study, which explains the positive MBA test after ozone gas treatment in most of them, a reduction of the toxin contents of mussel tissue was observed in the majority of samples, irrespective of mussel sample type (whole shucked or homogenized tissue), after ozone gas treatment (Table 2). This reduction was in the range of 6-100%, 25-83%, and 21-66% for free OA, OA esters and total OA, respectively. Practically, this means that in a worst case scenario (total OA reduction of 21%), non marketable mussels with total OA contents of 193 μg OA eq/kg, can be marketed after treatment with ozone gas, as toxin content would decrease to levels lower than the current regulatory limit (160 μg OA eq/kg shellfish tissue). No apparent deterioration of appearance or texture was noticed after ozone treatment.

In the same context, Croci et al. (1994) studied the depuration of live toxic mussels in ozonated artificial marine water (ozone generator providing 360 mg O_3/hr). A clear reduction of toxicity levels and OA was observed after 3 days of treatment, but some of the samples were still toxic after 6 days of depuration. Similarly, Reboreda et al. (2010) kept live mussels in ozonated seawater (Redox > 450 mV) for 24 hr and reported that mussels were not significantly detoxified by this process.

An important trend was observed with regard to reduction of toxin content in treated samples, when mussel sample type was considered (Table 2 and Figure 7), with percent reduction of OA content in homogenized tissue being much higher in comparison to that observed in whole shucked mussels. Toxin content in homogenized tissue decreased after treatment by 25-100%, 39-83%, and 51-66% for free OA, OA esters and total OA, respectively, whereas in the case of whole shucked mussels these values were in the range of 6-21%, 25-57%, and 21-33%, respectively. This observed effect could be possibly attributed to the greater penetration of gas ozone in mussel tissues leading to a higher interaction with OA molecules due to the larger surface area of the homogenized sample.

It is interesting to note that in the present study percent reduction of OA esters after ozone treatment was by far higher than the respective value for free OA in both whole shucked mussels and homogenized tissue. This is in partial agreement with the findings of Reboreda et al. (2010), who reported an increase in OA and a decrease of DTX3 (OA esters) after ozone treatment. Such a finding was characterized by these authors as an “unexpected result”.

Table 2. Toxicity of samples determined by MBA and concentrations of okadaic acid and acyl derivatives by LC-MS/MS of mussels' tissue before and after decontamination by ozone treatment at 15 ppm for 6 hr.

Sample no	Sampling Region	Sampling Year	Mussel sample form	Before decontamination: MBA (mice deaths)	Before decontamination: LC-MS/MS Free OA (μg/g)	Before decontamination: LC-MS/MS OA esters (μg/g)	Before decontamination: LC-MS/MS Total OA (μg/g)	After decontamination: MBA (mice deaths)	After decontamination: LC-MS/MS Free OA (μg/g)	After decontamination: LC-MS/MS OA esters (μg/g)	After decontamination: LC-MS/MS Total OA (μg/g)	Free OA reduction (%)	OA esters reduction (%)	Total OA reduction (%)
2	Thessaloniki	2009	Whole shucked	Negative (- - -)	ND[a]	ND	ND	Negative (- - -)	ND	ND	ND	-	-	-
4	Fthiotida	2009	Whole shucked	Negative (- - -)	ND	ND	ND	Negative (- - -)	ND	ND	ND	-	-	-
5	Piraeus	2009	Whole shucked	Positive (+++)	5.96	6.84	12.80	Positive (+++)	4.50	5.58	10.08	18	25	21
6	Megara	2009	Whole shucked	Positive (+++)	2.26	2.68	4.94	Positive (+++)	2.12	1.28	3.40	6	52	31
12	Thessaloniki	2007	Homogenated tissue	Positive (+++)	1.08	2.24	2.32	Positive (+++)	0.78	0.82	1.60	28	63	52
13	Pieria	2007	Homogenated tissue	Positive (+++)	0.84	0.58	1.42	Positive (+++)	0.24	0.24	0.48	71	59	66
14	Pieria	2007	Homogenated tissue	Positive (+++)	1.42	1.70	3.12	Positive (+++)	1.03	0.32	1.35	27	81	57
15	Pieria	2007	Homogenated tissue	Positive (+++)	0.32	0.66	0.98	Positive (+++)	0.24	0.11	0.35	25	83	64
16	Imathia	2007	Homogenated tissue	Positive (+++)	1.26	1.70	2.96	Positive (+++)	0.84	0.62	1.46	33	64	51
17	Imathia	2007	Homogenated tissue	Positive (+++)	0.16	0.90	1.06	Positive (+++)	ND	0.36	0.36	100	60	66
18	Megara	2007	Homogenated tissue	Positive (+++)	2.10	3.26	5.36	Positive (+++)	0.24	2.00	2.24	89	39	58
19	Market	2007	Whole shucked	Negative (+ - -)	ND	ND	ND	Positive (+++)	0.32	0.38	0.70	-	-	-
20	Market	2007	Whole shucked	Positive (+++)	1.90	0.70	2.60	Positive (+++)	1.51	0.30	1.81	21	57	30
21	Thessaloniki	2009	Whole shucked	Positive (+++)	1.40	0.65	2.05	Positive (+++)	1.13	0.30	1.43	19	54	30

[a] *ND*: not detected

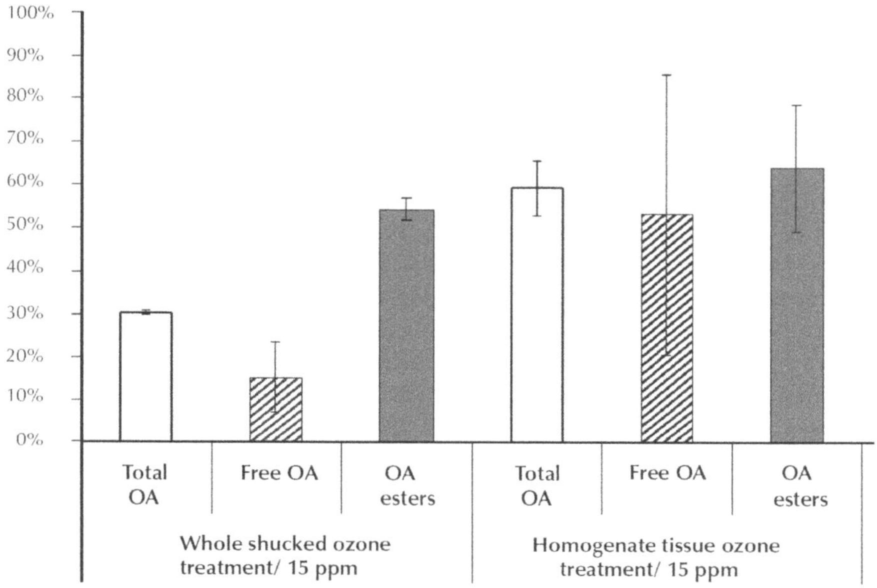

Figure 7. Mean percent reduction of DSP toxins (OA free form, OA esters and total OA) by ozone gas treatment in both shucked mussels and homogenated mussel tissue (n = 4 to 7). Error bars represent standard deviation of the mean.

Two of the negative samples, which were used as control samples for the treatment, remained MBA-negative following the ozone decontamination procedure with OA concentrations being still non-detectable, whereas one negative MBA sample (with one mouse dead in the MBA test) provided a positive MBA test after treatment (sample no. 19). This somewhat odd result could be attributed to one or both of the following factors: (a) inhomogeneity of the initial sample due to sample origin, as market samples normally consist of a mixture of mussels coming from different sampling locations and (b) a possible toxin transformation, which has been also implied by Reboreda et al. (2010); although in their study ozone did not substantially detoxify mussels, there was a modification in the ratio OA/Total toxin, whereas an increase in OA levels parallel to a decrease in DTX-3 (OA esters) levels was also observed in ozonated mussels, suggesting that the oxidative power of ozone could have been responsible for changing the mussels' toxin profile.

A possible explanation for the effect of ozone treatment on OA content of mussel tissue could be an interaction of gas ozone with the double bonds present in the OA molecule (Soriano et al., 2003), either in its free or esterified form. Ozone is known to attack double bonds of organic compounds (Smith, 2004) and OA molecule (both free and in esterified form) contains several double bonds which could be a potential target for ozone. Alteration of the free OA structure by ozone treatment could result in a reduction of its concentration because an altered molecule could possibly not be identifiable as OA in LC-MS/MS analysis and furthermore could be less toxic in the MBA. In this context, ozone treatment could also have contributed to a transformation of OA esters back to OA free form, which could account for the lower percent reduction of free OA observed in the present study, as some of the remaining free OA after processing could be derived from the esterified form. Further research is required to elucidate the actual mechanism of action of ozone on OA molecule.

CONCLUSION

This is the first report of the unambiguous identification of OA, DTX-1 and their esters in Greek mussels using LC-MS/MS. Present findings support the suitability of the LC-MS/MS for the determination of individual DSP toxins including ester derivatives as the results of MBA and LC-MS/MS methods showed very good agreement. Comparison of the MBA results with those of HPLC-FLD method, also, showed very good agreement. Further research on esterification of DSP toxins is required to verify if present findings are extendable to other shellfish from other locations.

Also, results of the present study indicate the potential for the commercialization of ozone treatment to decontaminate shucked mussels from DSP toxins. Taking into account that mussels are probably one of the best indicator species for evaluating the contamination of shellfish by toxins of the OA group, this processing method could be also applied to other contaminated shellfish species provided that they are marketed as shelled products. Further, research is underway to establish the optimal processing time/ozone dose combinations in order to achieve the best possible decontamination effects.

KEYWORDS

- **Diarrhetic shellfish poisoning;**
- **Dinophysistoxins**
- **Esters of okadaic acid**
- **Mussels**
- **Mouse bioassay**
- **Okadaic acid**

ACKNOWLEDGMENTS

The authors would like to thank the Mass Spectrometry Unit of the University of Ioannina for the providing access to LC-MS/MS facilities. Thanks are also expressed to all the relevant prefectural veterinary services for their contribution to the samplings. The collaboration of all staff members of the National Reference Laboratory of Marine Biotoxins is also greatly appreciated.

Chapter 2

Effect of Light on Vitamin Loss and Shelf Life of Pasteurized Milk Packaged in PET Bottles

Michael G. Kontominas, Anastasia V. Badeka, Nikolaos Pournis, and Ioannis Karabagias

INTRODUCTION

The losses in vitamins A, E, and B_2 as well as sensorial changes (odor, taste) in bactofuged, whole pasteurized milk stored at 4°C under fluorescent light and in the dark were monitored over a 15 day storage period. Milk containers tested included: 1 l bottles of (a) clear PET + UV blocker, 350-500 μm in thickness, (b) white colored PET + UV blocker, 350-400 μm in thickness, and (c) transparent blue colored PET + UV blocker, 350-400 μm in thickness. Milk packaged in 1 l coated paperboard cartons and stored under the same experimental conditions served as the "commercial control" sample. Based on sensory analysis, the shelf life of whole pasteurized milk tested in the present study was 13 days for samples packaged in clear blue PET + UV blocker bottles in the dark and 10 days under fluorescent light. The shelf life of milk packaged in the other three packaging materials was 10-11 days in the dark and 8-9 days under fluorescent light.

The losses for vitamin E were the highest followed by those for riboflavin and vitamin A. Average losses for vitamin A after 10 days of storage were 22.3%, 19.6%, 14.3%, and 22.3% in the dark and 36.6%, 28.6%, 23.3%, and 31.2% under fluorescent for samples packaged in clear PET + UV, white colored PET + UV, transparent blue colored PET + UV and control samples respectively. Respective losses were 41.2%, 29.4%, 23.9%, and 31% in the dark and 46.7%, 40.4%, 27.8%, and 39.6% under fluorescent light for vitamin E and 33.7%, 30.6%, 17.8%, and 30.2 % in the dark and 38.8%, 33.3%, 25.6%, and 31.8% under fluorescent for riboflavin.

Based on spectra transmission curves of packaging materials tested, it is suggested that the use of dark color pigmentation such as blue in fresh milk packaging will provide a better protection to light sensitive vitamins in cases where the expected shelf life of milk exceeds 5-6 days.

Modern milk producing/processing practices in Greece include collection of milk from selected dairy farms maintaining high hygienic standards and application of bactofugation in the processing plant that is the application of centrifugal force to remove a large percentage of microbial cells spores from milk prior to pasteurization. The number of microorganisms in raw milk may thus be reduced to 20,000-30,000 cfu/ml. Subsequent pasteurization may provide milk with a microbial load of 2,000-3,000 cfu/ml as opposed to 30,000 cfu/ml allowed by legislation (Papachristou et al., 2006a). This product has a refrigerated shelf life substantially longer than 5 days, which is the expected shelf life of pasteurized milk in Greece today. Given the contemporary trend to increase shelf life of pasteurized milk in Greece, the question rises as to what extent deterioration in nutritive (vitamin) content and sensory attributes of the product occur under conditions of commercial lighting over extended periods of time. Commercial lighting conditions during storage of milk include (a) glass door display cabinets or supermarket open refrigerators lit with fluorescent lamps and (b) house hold refrigerators maintaining the product in the absence of light (Papachristou et al., 2006a, 2006b).

Milk quality deterioration is perceived by the consumer through off-flavors that may be caused by chemical, physicochemical or microbiological changes in the product (Allen and Joseph, 1985; Rysstad et al., 1998; Thomas, 1981; Valero et al., 2000; Van Aardt et al., 2001). Among these defects, light induced off-flavors are the most common in milk attributed to two distinct causes. The first, a "burnt sunlight flavor" develops during the first 2 or 3 days of storage and is caused by degradation of sulfur containing amino acids of the whey proteins such as methionine to methional (Marsili, 1999). The second is a metallic or cardboardic off-flavor (lack of freshness) that develops 2 days later and does not dissipate. This off-flavor is attributed to lipid oxidation (Barnard, 1972). Light exposure especially to wavelengths below 520 nm also causes destruction of light-sensitive vitamins, mainly riboflavin, vitamins A and E (Bosset et al., 1994; DeMan, 1978, 1983; Fanelli et al., 1985; Hoskin, 1988; Hoskin and Dimick, 1979; Moyssiadi et al., 2004; Papachristou et al., 2006a, 2006b; Sattar et al., 1977; Skibsted, 2000; Vassila et al., 2002).

The packaging can directly influence the development of light-induced flavor by protecting the product from both light and oxygen (Moyssiadi et al., 2004; Papachristou et al., 2006a; Schröder, 1982; Skibsted, 2000; Van Aardt et al., 2001; Zygoura et al., 2004). Apart from traditional glass bottles and coated paperboard cartons, all-plastics containers, namely: high density polyethylene (HDPE), polycarbonate (PC), polyethylene terephthalate (PET) have been used in pasteurized milk packaging (Defosse, 2000; Papachristou et al., 2006b; Schröder et al., 1985). Problems with all-plastics containers used in numerous studies mentioned above, include transmission to light and permeability to oxygen. It should be noted, however, that oxidative reactions were reported to take place in milk packaged even in coated paperboard cartons which were found to be more or less permeable to oxygen (Rysstad et al., 1998; Schröder et al., 1985). More recently, PET bottles either clear with or without UV blocker or pigmented (usually with TiO_2) have been commercially used for fresh milk packaging (Bakish and Hatfield, 1995; Cladman et al., 1998; Moyssiadi et al., 2004; Papachristou et al., 2006a, 2006b; Van Aardt et al., 2001; Zygoura et al., 2004). The PET has excellent mechanical properties, is a good barrier to O_2 and reduces the adverse effects of light on milk quality in the form of pigmented bottles. The PET bottles provide excellent convenience and protection through easy opening and reclosing, thus minimizing recontamination (Cladman et al., 1998).

Most studies on milk shelf life determination involve keeping milk under refrigerated storage and fluorescent light. In certain cases premium quality pasteurized milk such as that described previously, remains on the supermarket shelve for 1 day while after purchase it is kept in the home refrigerator in the dark usually for another 3-4 days until consumed.

Given that bactofugation substantially increases milk shelf life, the objective of the present study was to determine (i) losses in vitamins A, B_2, and E in bactofuged pasteurized whole milk (ii) product shelf life as a function of packaging material with different light barriers under commercial lighting conditions of storage.

MATERIALS AND METHODS

Packaging Materials

Four different packages were evaluated with regard to their potential effect on retention of vitamins in milk during refrigerated storage. The four packages included: (a) a 1 l clear PET bottle + UV blocker, 350-400 μm in thickness, (b) a 1 l white colored PET bottle + UV blocker, 350-400 μm in thickness, (c) a 1 l clear blue colored PET bottle + UV blocker, 350-400 μm in thickness and (d) a 1 l PE coated paperboard carton (PE/paperboard/PE), 450 μm in thickness used as the "commercial control" sample. The PET bottle samples were produced by OLYMPOS S.A. (Larissa, Greece) on a model SBO 8, FIRIS 2 injection stretch blow molding line (SIDEL, France); coated paperboard cartons were also provided by OLYMPOS S.A. and produced by VARIOPACK (Homburg, Saar, Denmark).

Sample Preparation and Handling

Bactofuged whole milk (3.75% fat) was obtained from the dairy plant of OLYMPOS S.A. immediately after pasteurization. The milk was aseptically dispensed into the sterilized bottles and the coated paperboard cartons at the plant laboratory. Bottles were sealed using polypropylene screw caps while coated paperboard cartons were sealed on the production line sealer. Headspace for both cartons and bottles was between 25 and 40 ml. The filled containers were transported to the laboratory within 4 hr in polystyrene foam ice boxes packed in ice. Bottles were divided into two lots. The first lot was stored at 4 ± 0.5°C for a period of 24 hr under perpendicular fluorescent light provided by one 55 W cool white fluorescent lamp in a commercial display cabinet. The fluorescent lamp produced 825 ± 50 lux on the side surface of the containers, measured with a GE light meter (General Electric Co, USA). After this period samples were placed in a home type refrigerator in the dark (4°C) for an additional 14 days. The second lot was stored in the display cabinet described above for a period of 15 days. Milk samples (100 ml) were collected from sealed bottles at sampling times: 1, 3, 4, 6, 8, 10, 13, and 15 days after initial packaging for vitamin determination and sensorial testing. Testing on day 0 was carried out on milk samples immediately after packaging. On each sampling day four different samples (each from a different container) were assayed. After sampling each container was discarded.

Determination of Vitamins

Vitamins A and E were determined using the HPLC method of Zahar and Smith (1990). Riboflavin was determined using the HPLC method of Toyosaki, Yamamoto, and Mineshita (1988).

Physical Tests

Light Transmission Testing

Spectral transmission characteristics over the wavelength range of 350-780 nm (Rysstad et al., 1998) for all packaging materials were measured with a Shimadzu, model 2100 UV-VIS recording spectrophotometer.

Headspace Oxygen Measurement

A rubber septum (Systech Instr. Ltd, UK) was glued onto the surface of the bottle/carton and pierced with a 23 gauge needle connected to a pre-calibrated headspace analyzer model Gaspace 2 (Systech Instr. Ltd. UK) giving a direct reading of the percentage of oxygen in the bottle headspace.

Determination of Oxygen Transmission Rates

Oxygen transmission rates for all packages (bottles and paperboard cartons) were measured using the Oxtran 2/20 oxygen permeability tester (Mocon Controls, USA) at a relative humidity (RH) = 60%, temperature = 22°C and were expressed as ml/(package day atm). Whole bottles and cartons were epoxy glued to the special attachment of the oxygen permeability tester and a flow of nitrogen was established inside the container tested. Nitrogen was the carrier gas which swept the permanent atmospheric oxygen to the detector (coulometric detector).

Sensory Evaluation

A panel of 17 individuals consisting of faculty and graduate students, members of the Food Chemistry and Technology Laboratory in the Department of Chemistry were trained to differentiate between burnt (light oxidized) flavor and stale (lack of freshness) flavor of milk samples. Sensory data were collected between day 0 and 15 of storage. Milk samples (30 ml) were presented to panelists in individual sensory booths as sets of four with a resting interval of 1 or 2 min between sample sets. Panelists rated each sample for intensity and type of off-flavor on a scale between 0-5 where a score of 5 corresponded to very good flavor milk and 0 to unfit for consumption (very strong off-flavor) milk (IDF, 1987). A score of 3.5 was taken as the lower limit of acceptability.

Statistical Analysis

The experiment was replicated twice on different occasions. On each sampling day 3 different samples per treatment were analyzed ($n = 2 \times 3 = 6$). Data were subjected to analysis of variance using the Excel 97 software program (Microsoft, CA, USA) and where statistical differences were noted, differences among packages were determined, using the least significant difference (LSD) test. Significance was defined at $P < 0.05$.

DISCUSSION AND RESULTS

Vitamin A and E Retention

Vitamin A and E content of whole milk samples packaged in various PET containers as a function of storage time under both fluorescent light and in the dark are given in Tables 1 and 2. It is clear that for milk stored in the dark (Table 1) all packaging

Table 1. Retention of vitamins A and E (μg/ml) in whole, bactofuged, pasteurized milk packaged in various containers during storage at 4°C in the dark.

Packaging material	Days of storage																	
	0		1*		3		4		6		8		10		13		15	
	A	E	A	E	A	E	A	E	A	E	A	E	A	E	A	E	A	E
Clear PET + UV	0.56[a] ± 0.02	1.275[a] ± 0.05	0.53[a] ± 0.03	1.14[a] ± 0.06	0.51[a] ± 0.03	1.09[a] ± 0.07	0.49[a] ± 0.03	1.04[a] ± 0.06	0.47[a] ± 0.02	0.99[a] ± 0.05	0.45[a] ± 0.02	0.86[a] ± 0.04	0.435[a] ± 0.03	0.75[a] ± 0.03	0.42[a] ± 0.02	0.69[a] ± 0.04	0.40[a] ± 0.03	0.65[a] ± 0.03
White coloured PET + UV	0.56[a] ± 0.02	1.275[a] ± 0.05	0.52[a] ± 0.02	1.16[a] ± 0.05	0.50[a] ± 0.03	1.10[a] ± 0.06	0.48[a] ± 0.02	1.06[a] ± 0.05	0.48[a] ± 0.03	1.08[a] ± 0.04	0.46[a] ± 0.03	0.96[b] ± 0.05	0.45[a] ± 0.02	0.90[b] ± 0.04	0.43[a] ± 0.01	0.85[bc] ± 0.05	0.415[a] ± 0.02	0.745[b] ± 0.04
Clear blue colored PET + UV	0.56[a] ± 0.02	1.275[a] ± 0.05	0.54[a] ± 0.03	1.20[a] ± 0.05	0.52[a] ± 0.02	1.16[a] ± 0.05	0.51[a] ± 0.02	1.13[a] ± 0.05	0.51[a] ± 0.02	1.11[a] ± 0.07	0.49[a] ± 0.03	1.02[b] ± 0.04	0.48[a] ± 0.03	0.97[b] ± 0.04	0.46[a] ± 0.02	0.94[c] ± 0.04	0.44[a] ± 0.03	0.925[c] ± 0.05
Paperboard carton	0.56[a] ± 0.02	1.275[a] ± 0.05	0.53[a] ± 0.02	1.18[a] ± 0.04	0.51[a] ± 0.02	1.11[a] ± 0.07	0.48[a] ± 0.01	1.08[a] ± 0.04	0.47[a] ± 0.02	1.06[a] ± 0.05	0.44[a] ± 0.02	0.92[ab] ± 0.05	0.435[a] ± 0.01	0.88[b] ± 0.05	0.42[a] ± 0.03	0.82[b] ± 0.03	0.40[a] ± 0.02	0.755[b] ± 0.04

Values reported are the mean of six determinations (n = 2 × 3 = 6).

* First day under fluorescent light, rest 14 days in the dark.

a, b, ... Different letters in columns represent statistically significant differences ($P < 0.05$).

Table 2. Retention of vitamins A and E (μg/ml) in whole, bactofuged, pasteurized milk packaged in various containers during storage at 4°C under fluorescent light.

Packaging material	Days of storage																	
	0		1		3		4		6		8		10		13		15	
	A	E	A	E	A	E	A	E	A	E	A	E	A	E	A	E	A	E
Clear PET + UV	0.56^{a}	1.275^{a}	0.50^{a}	1.13^{a}	0.45^{a}	0.96^{a}	0.41^{a}	0.92^{a}	0.38^{a}	0.75^{a}	0.36^{a}	0.69^{a}	0.355^{a} ± 0.02	0.68^{a}	0.29^{a}	0.68^{a}	0.255^{a}	0.53^{a}
	± 0.02	± 0.05	± 0.03	± 0.06	± 0.02	± 0.07	± 0.03	± 0.06	± 0.02	± 0.04	± 0.02	± 0.03		± 0.04	± 0.02	± 0.02	± 0.01	± 0.02
White coloured PET +UV	0.56^{a}	1.275^{a}	0.54^{a}	1.18^{a}	0.49^{a}	1.04^{a}	0.45^{a}	1.02^{a}	0.43^{ab}	0.86^{b}	0.42^{ab}	0.74^{a}	0.40^{b}	0.76^{ab}	0.38^{b}	0.69^{a}	0.365^{b}	0.62^{b}
	± 0.02	± 0.05	± 0.02	± 0.07	± 0.03	± 0.05	± 0.02	± 0.04	± 0.03	± 0.04	± 0.03	± 0.04	± 0.01	± 0.04	± 0.01	± 0.03	± 0.02	± 0.03
Clear blue colored PET + UV	0.56^{a}	1.275^{a}	0.53^{a}	1.19^{a}	0.51^{a}	1.10^{a}	0.48^{a}	1.06^{a}	0.46^{b}	1.03^{c}	0.45^{b}	0.98^{c}	0.43^{b}	0.92^{c}	0.40^{b}	0.89^{c}	0.39^{b}	0.85^{c}
	± 0.02	± 0.05	± 0.03	± 0.05	± 0.04	± 0.10	± 0.04	± 0.08	± 0.03	± 0.06	± 0.02	± 0.04	± 0.02	± 0.06	± 0.03	± 0.04	± 0.03	± 0.04
Paperboard carton	0.56^{a}	1.275^{a}	0.52^{a}	1.17^{a}	0.48^{a}	1.09^{a}	0.44^{a}	0.98^{a}	0.41^{ab}	0.96^{c}	0.40^{ab}	0.86^{b}	0.385^{a} ± 0.01	0.77^{b}	0.35^{b}	0.78^{b}	0.35^{b}	0.67^{b}
	± 0.02	± 0.05	± 0.04	± 0.05	± 0.02	± 0.09	± 0.03	± 0.04	± 0.04	± 0.04	± 0.03	± 0.03		± 0.03	± 0.02	± 0.02	± 0.02	± 0.03

Values reported are the mean of six determinations (n = 2 x 3 = 6).

$^{a, b, \ldots}$ Different letters in columns represent statistically significant differences ($P < 0.05$)

materials including the control resulted in the same losses of vitamin A (loss equal to 21.4-28.6% on day 15 of storage). In contrast, vitamin E losses varied significantly ($P < 0.05$) among packaging materials starting with day 8 of storage with the greater retention in vitamin E provided by the clear blue PET + UV container. The losses of vitamin E recorded for the four different containers were equal to 27.4%, 40.8%, 41.6%, and 49% for the clear blue PET + UV, the paperboard carton, the white colored PET + UV and the clear PET + UV container respectively on day 15 of storage.

Regarding milk samples stored under fluorescent light (Table 2), statistically significant ($P < 0.05$) differences in retention of vitamin A were recorded among different packaging treatments starting with day 6 of storage. That is, the clear blue PET + UV, white colored PET + UV and the paperboard carton provided similar protection to vitamin A but higher compared to the clear PET + UV packaging material. On day 15 of storage losses of vitamin A recorded for the four different containers were equal to 30.4%, 37.5%, 34.8%, and 54.5% for the clear blue PET + UV, the paperboard carton, the white colored PET + UV and the clear PET + UV container respectively.

Based on O_2 transmission values (Table 3), headspace volume (between 25-40 ml) and headspace oxygen concentration values between 21 and 8% on day 15 of storage (data not shown), it can be postulated that within the relatively short life span of the experiment, oxygen was not a major factor in vitamin A degradation. Oxygen transmission rates of paperboard cartons being almost 40 times higher than those of PET containers support this statement, since this type of container provided a better protection to vitamins, including vitamin A than clear PET.

Table 3. Oxygen transmission rate of packaging materials.

Packaging material	O_2 transmission rate* M (package×day×atm)$^{-1}$
Clear PET + UV	$0.90^{a}\pm0.08$
White colored PET + UV	$0.85^{a}\pm0.07$
Clear blue colored PET + UV	$0.78^{a}\pm0.08$
Paperboard carton	$33.6^{b}\pm2.8$

Values reported are the mean of six determinations (n = 2 × 3 = 6).
* T = 22°C, RH = 60%
[a, b] Different letters in columns represent statistically significant differences ($P < 0.05$)

Statistically significant ($P < 0.05$) differences among the various packaging treatments were recorded for vitamin E starting with day 6 of storage. That is, the clear blue PET + UV container provided the best protection to vitamin E, followed by the paperboard carton and the white colored PET + UV container (showing similar protection) and lastly by the clear PET + UV container providing the least protection to vitamin E. On day 15 of storage, losses of vitamin E recorded for the four different containers were equal to 33.3%, 47.4%, 51.4%, and 58.4% for the clear blue PET + UV, the paperboard carton, the white colored PET + UV and the clear PET + UV container respectively. However, vitamin E proved to be more sensitive to degradation than vitamin A. Present data on vitamin A and E are in general agreement with those of the literature given the differences in experimental conditions and types of packaging materials used. Fanelli et al. (1985) reported losses of vitamin A between 6.3 and 50% after milk irradiation for 24 hr, with the best protection being achieved through the use of special UV blockers compounded into the HDPE resin used to manufacture plastic bottles. DeMan (1978) exposed whole, semi skimmed (2% fat) and skimmed milk to a fluorescent light intensity of 2200 lux for up to 48 hr at refrigerator temperature. Vitamin A of whole milk dropped to 67.7% of its original content after 30 hr and remained constant for a further 18 hr. In 2% milk it dropped to 23.6% and in skim milk to 4.2% of its original content. Gaylord et al., (1986) studied the effect of solids (NEDM); light intensity and fat content on vitamin content in milk packaged in glass test tubes and reported a 57%, 53%, and 45% retention of retinol palmitate in whole, 2% and skim milk respectively after 48 hr of fluorescent light exposure. Desarzens et al., (1983) exposed whole unpacked pasteurized milk to cool white fluorescent light and reported a 50% degradation of vitamin A after only 6 hr. A strong decrease in vitamin E concentration was also observed. Sattar et al. (1977) showed that vitamin A content of milk decreased at a wavelength below 415 nm and to a lesser extent at wavelengths below 415-455 nm. Cladman et al. (1998) studied vitamin A degradation in whole milk as a function of packaging material light transmittance and reported a decrease by at least 47% in clear PET bottles and LDPE pouches after 6 days of storage. A significantly lower decrease in vitamin A (28%) was observed in green colored PET bottles while sample in HDPE jugs suffered the most severe vitamin A losses (58%). Vassila et al. (2002) similarly reported losses in vitamin A in whole milk between 15.1 and 73.0% in various flexible pouch materials. Moyssiadi et al. (2004) reported vitamin A losses in low fat milk equal to 11% for pigmented PET bottles, 16% for paperboard cartons, and 31% for clear PET bottles after 7 days of storage. Zygoura et al. (2004) reported respective vitamin A losses for whole milk equal to 29.8%, 14.0%, and 50.9% after 7 days of storage under fluorescent light using the same as above packaging materials. Finally, Papachristou et al. (2006a) reported losses of vitamin A equal to 18.2%, 27.0%, and 17.5% and respective losses of vitamin E equal to 42.7%, 53.6%, and 43.9% after 10 days of storage under fluorescent light in clear PET + UV, clear PET and paperboard carton respectively.

A final important point to be made is that the two main mechanisms of milk quality deterioration are chemical oxidation through O_2 permeation and light-induced oxidation/vitamin degradation. The chemical nature and thickness (350-400μm) of the PET packaging material may effectively control the first mechanism (given a reasonably small headspace), while UV absorbers and pigmentation may control the latter.

Riboflavin

The destruction of riboflavin in milk is directly related to the radiant energy emitted between wavelengths 400 and 520 nm. Riboflavin content of whole milk samples in various containers as a function of storage time both in the dark and under fluorescent light are given in Tables 4 and 5. Statistically significant ($P < 0.05$) differences among packaging treatments were recorded for milk samples stored both under fluorescent

Table 4. Retention of riboflavin (µg/ml) in whole, bactofuged, pasteurized milk packaged in various containers during storage at 4°C in the dark.

Packaging material	Days of storage								
	0	**1***	**3**	**4**	**6**	**8**	**10**	**13**	**15**
Clear PET + UV	$1.29^{a}\pm0.06$	$1.14^{a}\pm0.06$	$1.04^{a}\pm0.05$	$0.99^{a}\pm0.06$	$0.94^{a}\pm0.06$	$0.89^{a}\pm0.04$	$0.855^{a}\pm0.04$	$0.81^{a}\pm0.03$	$0.76^{a}\pm0.04$
White colored PET + UV	$1.29^{a}\pm0.06$	$1.14^{a}\pm0.05$	$1.08^{ab}\pm0.06$	$1.02^{a}\pm0.07$	$0.97^{a}\pm0.07$	$0.93^{a}\pm0.05$	$0.895^{a}\pm0.03$	$0.86^{a}\pm0.04$	$0.85^{a}\pm0.05$
Clear blue colored PET + UV	$1.29^{a}\pm0.06$	$1.22^{a}\pm0.07$	$1.18^{b}\pm0.04$	$1.15^{a}\pm0.09$	$1.13^{b}\pm0.08$	$1.08^{ab}\pm0.04$	$1.06^{b}\pm0.05$	$1.01^{b}\pm0.03$	$0.96^{b}\pm0.05$
Paperboard carton	$1.29^{a}\pm0.06$	$1.20^{a}\pm0.04$	$1.14^{ab}\pm0.05$	$1.10^{a}\pm0.08$	$1.06^{ab}\pm0.06$	$0.99^{a}\pm0.06$	$0.90^{a}\pm0.04$	$0.88^{a}\pm0.05$	$0.84^{a}\pm0.04$

Values reported are the mean of six determinations (n = 2 x 3 = 6).

* First day under fluorescent light, rest 14 days in the dark.

$^{a, b, \ldots}$ Different letters in columns represent statistically significant differences ($P < 0.05$)

Table 5. Retention of riboflavin (µg/ml) in whole, bactofuged, pasteurized milk packaged in various containers at 4°C under fluorescent light.

Packaging material	Days of storage								
	0	1	3	4	6	8	10	13	15
Cleat PET + UV	$1.29^{a}\pm0.06$	$1.15^{a}\pm0.06$	$1.01^{a}\pm0.05$	$0.98^{a}\pm0.03$	$0.93^{a}\pm0.04$	$0.86^{a}\pm0.05$	$0.79^{a}\pm0.05$	$0.72^{a}\pm0.04$	$0.67^{a}\pm0.05$
White colored PET + UV	$1.29^{a}\pm0.06$	$1.16^{a}\pm0.05$	$1.05^{a}\pm0.04$	$1.00^{a}\pm0.04$	$0.96^{a}\pm0.05$	$0.91^{ab}\pm0.05$	$0.86^{a}\pm0.06$	$0.83^{a}\pm0.05$	$0.80^{b}\pm0.04$
Clear blue colored PET + UV	$1.29^{a}\pm0.06$	$1.21^{a}\pm0.07$	$1.10^{a}\pm0.06$	$1.05^{a}\pm0.04$	$1.02^{b}\pm0.04$	$0.99^{b}\pm0.05$	$0.96^{b}\pm0.03$	$0.94^{b}\pm0.03$	$0.93^{c}\pm0.06$
Paperboard carton	$1.29^{a}\pm0.06$	$1.19^{a}\pm0.04$	$1.08^{a}\pm0.05$	$1.04^{a}\pm0.05$	$1.01^{ab}\pm0.05$	$0.93^{ab}\pm0.04$	$0.88^{a}\pm0.04$	$0.85^{ab}\pm0.05$	$0.76^{ab}\pm0.04$

Values reported are the mean of six determinations ($n = 2 \times 3 = 6$).

[a, b, ...] Different letters in columns represent statistically significant differences ($P < 0.05$)

light and in the dark starting with day 6 of storage. The clear blue PET + UV container provided the best protection to vitamin B_2. The losses of riboflavin after 15 days of storage were 26.4%, 34.9%, 34.1%, and 41.1% in the dark and 27.9%, 41.1%, 38%, and 48.1% under fluorescent light for clear blue PET + UV, paperboard carton, white colored PET + UV and clear PET + UV container respectively. Present data on vitamin B_2 are in general agreement with those in the literature given the differences in experimental conditions and the type of packaging materials used. Fanelli et al. (1985) reported losses of riboflavin between 57 and 70% after 16 hr of irradiation using a variety of pigments and UV absorbers in their HDPE containers.

De Man (1983) reported riboflavin losses, after exposure of milk containers to fluorescent light for 48 hr, equal to 16.6% for paperboard cartons, 28.4% for clear PET pouches, 18.8% for HDPE jugs, and 15.3% for jugs pigmented with 2% TiO_2. Hoskin and Dimick (1979) reported riboflavin losses after exposure of milk containers to fluorescent light for 72 hr, equal to 13% for clear PC bottles, 10% HDPE bottles, 10% for paperboard carton, and 6% for PC tinted bottles. No significant loss of riboflavin was observed in milk held in the dark. Gaylord et al. (1986) studied the destruction of riboflavin in milk and reported 45%, 36%, and 30% retention in vitamin B_2 for whole, 2% and skimmed milk respectively after 48 hr of fluorescent light exposure. Hoskin (1988) studied riboflavin degradation in whole milk packaged in shielded with aluminum foil and unshielded half gallon HDPE containers. Control samples (not exposed to light) had a riboflavin content of 0.921 μg/ml after 72 hr of light exposure. Respective contents for unshielded and aluminum foil shielded containers were 0.730 and 0.844 μg/ml. Desarzens et al. (1983) exposed unpackaged pasteurized milk to cool white fluorescent light and reported riboflavin losses over 75% after 7 hr. Vassila et al. (2002) reported losses in vitamin B_2 between 18.8 and 45.3% for whole milk after storage for 7 days in various plastic pouch materials. Moyssiadi et al. (2004) reported riboflavin losses in low fat milk equal to 33% for pigmented PET bottles, 28% for paperboard cartons and 40% for clear PET bottles after 7 day storage riboflavin. Zygoura et al. (2004) reported respective riboflavin losses for whole milk equal to 30.9, 19.8, and 47.1% after 7 days of storage using the same as above packaging materials. Finally, Papachristou et al. (2006a) reported riboflavin losses equal to 38.7%, 52.5%, and 35.0% for clear PET + UV, clear PET and paperboard cartons.

Sensory Evaluation

Flavor evaluation results are given in Tables 6 and 7 for milk samples stored in the dark and under fluorescent light respectively. Based on data in Table 6, milk stored in the dark retained its flavor better than that stored under fluorescent light. It is also clear that off-flavors increased more rapidly in milk packaged in light exposed containers (clear PET + UV) versus light protected containers (clear blue PET + UV, white colored PET + UV and paperboard carton). Milk stored in the dark reached its acceptability limit (score: 3.5) after 13 days for clear blue PET + UV and 10 days for the other of packaging materials. Milk stored under fluorescent light reached the same limit after 10 days for clear blue PET + UV and after 8 days for the other packaging materials. Flavor defects were described as "slightly stale", to "stale", "plastic taste oxidized" and "sour".

Table 6. Flavor evaluation of whole, bactofuged, pasteurized milk packaged in various containers during storage at 4°C in the dark.

Packaging material	Days of storage								
	0	1*	3	4	6	8	10	13	15
Clear PET + UV	$5^a±0.4$	$5^a±0.3$	$4.9^a±0.3$	$4.8^a±0.3$	$4.5^a±0.4$	$4.1^a±0.2$	$4.0^a±0.3$	$3^a±0.2$ stale	$2^a±0.1$ Sour
White colored PET + UV	$5^a±0.4$	$5^a±0.4$	$5^a±0.3$	$4.8^a±0.4$	$4.7^a±0.3$	$4.4^a±0.3$	$4.1^a±0.2$	$3.2^a±0.3$ slightly stale	$2.8^c±0.2$ stale
Clear blue colored PET + UV	$5^a±0.4$	$5^a±0.5$	$5^a±0.4$	$4.9^a±0.3$	$4.7^a±0.5$	$4.6^a±0.4$	$4.3^a±0.3$	$3.5^a±0.2$	$3.2^c±0.2$ slightly stale
Paperboard carton	$5^a±0.4$	$5^a±0.3$	$5^a±0.3$	$4.8^a±0.4$	$4.4^a±0.2$	$4.2^a±0.4$	$3.8^a±0.2$	$3.2^a±0.2$ slightly stale	$2.4^b±0.1$ sour

Results are expressed as scores/comments. Numerical scale of scoring: very good = 5, good = 4, fair = 3, poor = 2, very poor = 1, unfit of consumption = 0.

* First day under fluorescent light, rest 14 days in the dark.

[a, b, ...] Different letters in columns represent statistically significant differences ($P < 0.05$)

Table 7. Flavor evaluation of whole, bactofuged, pasteurized milk packaged in various containers during storage at 4°C under fluorescent light.

Packaging material	Days of storage								
	0	**1**	**3**	**4**	**6**	**8**	**10**	**13**	**15**
Clear PET + UV	$5^{a}\pm0.4$	$5^{a}\pm0.2$	$4.6^{a}\pm0.3$	$4.4^{a}\pm0.4$	$4.2^{a}\pm0.3$	$3.7^{a}\pm0.3$	$3.4^{a}\pm0.2$	$2.8^{a}\pm0.2$	$1.6^{a}\pm0.1$
							slightly stale	plastic taste oxidized	Sour
White colored PET + UV	$5^{a}\pm0.4$	$5^{a}\pm0.3$	$5^{a}\pm0.3$	$4.8^{a}\pm0.3$	$4.7^{a}\pm0.4$	$4.0^{ab}\pm0.3$	$3.4^{a}\pm0.3$	$3.1^{ab}\pm0.3$	$1.9^{ab}\pm0.2$
							slightly stale		
Clear blue colored PET + UV	$5^{a}\pm0.4$	$5^{a}\pm0.3$	$5^{a}\pm0.2$	$4.8^{a}\pm0.4$	$4.7^{a}\pm0.5$	$4.5^{b}\pm0.4$	$4.2^{b}\pm0.3$	$3.4^{b}\pm0.3$	$2.8^{c}\pm0.2$
								slightly stale	
Paperboard carton	$5^{a}\pm0.4$	$5^{a}\pm0.4$	$5^{a}\pm0.4$	$4.8^{a}\pm0.3$	$4.2^{a}\pm0.3$	$3.9^{ab}\pm0.3$	$3.4^{a}\pm0.2$	$2.5^{a}\pm0.2$	$2.0^{b}\pm0.1$
								slightly sour	sour

Results are expressed as scores/comments. Numerical scale of scoring: very good = 5, good = 4, fair = 3, poor = 2, very poor = 1, unfit of consumption = 0.
[a, b, ...] Different letters in columns represent statistically significant differences ($P < 0.05$)

The present sensory results are in agreement with those of van Aardt et al. (2001) who reported better milk quality retention in amber PET and UV compounded PET vs. clear PET bottles; and those of Cladman et al. (1998) who reported superior milk protection against light oxidation when packaged in green PET versus clear PET bottles. Both Moyssiadi et al. (2004) and Zygoura et al. (2004) reported better milk (full fat and low fat) quality protection in pigmented (with TiO_2) PET bottles and paperboard carton as compared to clear PET bottles after 5 days of storage. After 7 days of storage milk packaged in all above packaging materials was unfit for consumption. Finally, Papachristou et al. (2006a) reported slightly better flavor retention of full fat milk packaged in clear PET + UV than that packages in clear PET and paperboard carton.

CONCLUSION

Bottles made of clear blue PET + UV provided better protection with regard to retention of vitamins B_2, A, and E as compared to the white colored PET + UV and the paperboard carton while clear PET + UV provided the least amount of protection of milk against vitamin A, E, and B_2 degradation. Based on sensory evaluation the shelf life of whole bactofuged, pasteurized milk packaged in clear blue PET + UV is 13 days in the dark and 10 days under fluorescent light. Respective shelf life of milk packaged in the other three packaging materials is 10-11 days in the dark and 8-9 days under fluorescent light. After an average storage of 10 days of storage, losses in riboflavin in clear blue PET + UV, white colored PET + UV, paperboard carton and clear PET + UV were 17.9%, 30.6%, 30.2%, and 33.7% for milk stored in the dark and 25.6%, 33.3%, 31.8%, and 38.8% for milk stored under fluorescent light. For vitamin A, losses were respectively 14.3%, 19.6%, 22.3%, and 22.3% in the dark and 23.2%, 28.6%, 31.2%, and 36.6% under fluorescent light. Finally for vitamin E losses were respectively 23.9%, 29.4%, 31.0%, and 41.2% in the dark and 27.8%, 40.4%, 39.6%, and 46.7% under fluorescent light.

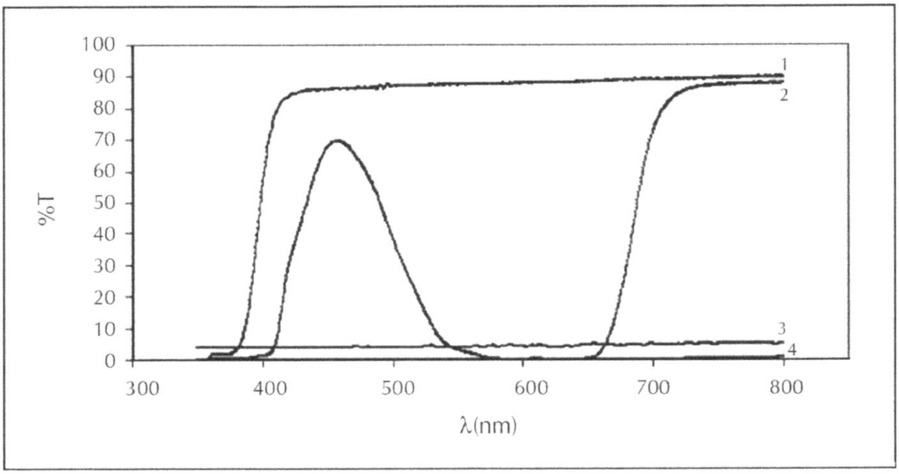

Figure 1. Figure 1. Spectral-transmission curves of various milk packaging materials. Clear PET + UV blocker (1), PET + UV blocker (2), coated paperboard carton (3) PET + UV blocker.

Based on spectral transmission curves of packaging materials tested (Figure 1), it is suggested that the use of a UV blocking agent in combination with a dark color pigmentation (i.e. blue) in PET milk bottles will provide a refrigerated shelf life of roughly 13 days in the dark and 10 days under fluorescent light with the least amount of vitamin losses as compared to other light protected containers. The PET packaging with such characteristics is proposed as an excellent alternative to the commercial paperboard carton for the packaging of pasteurized milk.

KEYWORDS

- **Light transmission**
- **Milk**
- **PET packaging**
- **Shelf life**
- **Vitamins**

ACKNOWLEDGMENTS

The authors would like to thank OLYMPOS S.A. for providing the milk and packaging samples.

Chapter 3

Differentiation of Greek Extra Virgin Olive Oils According to Cultivar and Geographical Origin

D. Alevras, M. Tasioula-Margari, K. Karakostas, A. Sacco, and M. G. Kontominas

INRTODUCTION

The aim of the present study was to classify extra virgin olive oils (EVOOs) from Western Greece according to cultivar and geographical origin, based primarily on triglyceride (TG) composition and secondarily on pigment content (chlorophylls and carotenoids), by means of Linear Discriminant Analysis. A total of 96 olive oil samples were collected during the harvesting periods 2006–2007 and 2007–2008 (48 samples × 2 harvesting periods), from six regions of Western Greece and from five local cultivars. The analysis of TGs was performed by high performance liquid chromatography (HPLC). Thirty one different TGs were determined. Using the TG composition data, the olive oil samples were sufficiently classified according to cultivar (89.6%) and geographical origin (75.0%). Combination of TGs data with pigments data resulted in a 91.7% and 83.3% classification, respectively. The olive oil samples of the Koroneiki cultivar were 83.3% correctly classified according to geographical origin, using the TG composition, or TG composition plus pigment content as variables.

The European Union (EU) is the major olive oil producer in the world, with 75% of the total production. Within the EU, Spain (43%), Italy (32%), and Greece (22%) provide 97% of European olive oil production (Ollivier et al., 2003). Greece produces approximately 400,000 tons of olive oil annually (Petrakis et al., 2008).

Olive oil received prominent attention during the last decades, as it is a major constituent of the Mediterranean diet. Its constituents exhibit a protective effect against different types of cancer and significantly reduce mortality caused by heart disease (Owen et al., 2004). Given its value to health and its production cost, olive oil is more expensive than other types of oils, and is thus a target for adulteration (Lorenzo et al., 2002).

The EVOO has a highly variable chemical composition (Boggia et al., 2002). This variability mostly depends on olive cultivars, climatic conditions, and agricultural practices (Fontanazza et al., 1994; Gutierrez et al., 1999). These factors can affect the composition of TGs (Haddada et al., 2007; Stefanoudaki et al., 1997); the main components of olive oil accounting for approximately 95–98% by weight.

The need to determine the origin of an olive oil has become necessary after the introduction of the "Protected Designation of Origin" concept to olive oils (EC Regulation 1187/2000). Classification of an olive oil as a PDO product involves a thorough knowledge of the physical and chemical characteristics as well as properly defined cultivar names (Stefanoudaki et al., 1997). This designation guarantees that the quality of the product is closely linked to its territorial origin. Characterization of an olive oil as PDO increases the added value of the product and promotes its marketing in domestic and foreign markets.

Characterization of an olive oil related to a given geographical origin and olive fruit variety requires the processing of analytical data as a whole, using mathematical and statistical techniques (Bucci et al., 2002). Chemometric and statistical methods for the characterization, authentication, and classification of virgin olive oils have been applied to several chemical components, including TGs (Stefanoudaki et al., 1997); fatty acids (Di Bella et al., 2007), TGs and fatty acids (Bronzini de Caraffa et al., 2008); sterols (Temime et al., 2008); TGs and sterols (Diaz et al., 2005); fatty acids and sterols (Boggia et al., 2002); TGs, fatty acids and sterols (Haddada et al., 2007); TGs and volatile compounds (Oueslati et al., 2008); sterols, triterpenic alcohols and hydrocarbons (Ferreiro and Aparicio, 1992); volatile compounds (Luna et al., 2006) and so on.

The spectroscopic techniques, without previous sample treatment, have also been used for the characterization and classification of virgin olive oils including: 1H, 13C, and 31P-NMR spectroscopy (D'Imperio et al., 2007; Petrakis et al., 2008); NIR spectroscopy (Galtier et al., 2007); ICP (Benincasa et al., 2007); PTR-MS (Araghipour et al., 2008); spectral nephelometry (Mignani et al., 2005); excitation-emission fluorescence spectroscopy (Dupuy et al., 2005). Finally techniques such as IRMS (Bianchi et al., 1993); electronic nose (Cimato et al., 2006); and electronic nose in combination with artificial neural network (Cosio et al., 2006); have been used. Several studies have been carried out to correlate the chemical composition of olive oil to its geographic origin (Di Bella et al., 2007; Ferreiro and Aparicio, 1992; Ollivier et al., 2003; Petrakis et al., 2008; Temime et al., 2008; Tsimidou and Karakostas, 1993) or cultivar (Boschelle et al., 1994; Dhifi et al., 2005; Haddada et al., 2007; Stefanoudaki et al., 1999; Tura et al., 2008). Most of these studies refer to Spanish and Italian olive oil samples while only limited work has been carried out on Greek olive oil samples (Petrakis et al., 2008; Stefanoudaki et al., 1997, 1999; Tsimidou and Karakostas, 1993; Tsimidou et al., 1987).

Regarding classification of Greek olive oils most work has been done using basically the main Greek variety that is Koroneiki. Furthermore, regarding the geographical origin of olive oils no information exists on olive oils from Western Greece. Based on the above the objective of the present study was to differentiate Greek EVOOs according to cultivar and geographical origin from six regions of Western Greece (Zakynthos, Kefalonia, Lefkada, Kerkyra, Preveza, and Messologi) belonging to five local cultivars (Koroneiki, Ntopia of Zakynthos, Thiaki, Asprolia, and Lianolia). It should be noted that olive oil data from this part of Greece is scarce and that all olive orchards chosen lie within a radius of lower or equal to 150 km of each other having similar climatic conditions. Parameters determined included TGs, acidity, K_{232}, K_{270}, chlorophylls, and carotenoids. Experimental data were treated using ANOVA and multivariate statistical analysis (Linear Discriminant Analysis).

MATERIALS AND METHODS

Olive Oil Samples

A total of 96 EVOOs were collected during the harvesting periods 2006–2007 and 2007–2008 (48 samples × 2 harvesting periods), from six regions of Western Greece (Zakynthos, Kefalonia, Lefkada, Kerkyra, Preveza, and Messologi). The EVOO samples chosen for this study were from five local olive cultivars (Koroneiki, Ntopia of Zakynthos, Thiaki, Asprolia, and Lianolia). The sampling was carried out during the months November until the end of January. Olives were picked by hand at the stage of optimum maturity and processed in selected local olive mills, using the traditional three phase system technology (washing, de-leafing, crushing, malaxation at 30°C for 30 min). All olive oil samples were separated by centrifugation at room temperature and filtered. After filtration, the olive oils were stored in dark glass bottles at 4°C under a nitrogen atmosphere until chemical analyses were performed. Table 1 shows the number of extra virgin olive oil samples collected along with cultivar and geographical origin.

Table 1. The EVOO samples collected from Western Greece.

Origin	Cultivar	Number of samples[a]
Zakynthos	Koroneiki	8+8
Zakynthos	Ntopia of Zakynthos	6+6
Kefalonia	Koroneiki	7+7
Kefalonia	Thiaki	5+5
Lefkada	Asprolia	6+6
Lefkada	Koroneiki	3+3
Kerkyra	Lianolia	4+4
Kerkyra	Koroneiki	3+3
Preveza	Lianolia	3+3
Messologi	Koroneiki	3+3

[a] Represents the sun of samples collected during the harvesting periods 2006–2007 and 2007–2008

Determination of Conventional Quality Indices

Acidity and UV absorbance (K_{232}, K_{270}) were determined according to the European Official Methods of Analysis (EC Regulation 2568/91). Acidity, expressed as the percent of oleic acid, was determined volumetrically after the dissolution of olive oil in ethanol/ether (1:1) and titration with 0.1 N sodium hydroxide. The K_{232} and K_{270} specific extinction coefficients were calculated from the absorption value at 232 and 270 nm, respectively, of 1% solution of olive oil in isooctane using UV spectrophotometer (Model Lambda 25, PerkinElmer) and quartz cell thickness of 1 cm. All determinations were made in triplicate.

Determination of Chlorophyll and Carotenoid Content

The amounts of chlorophyll and carotenoid compounds were determined from the maximum of absorption of the olive oil sample dissolved in cyclohexane at 670 and

470 nm, respectively, using the specific extinction coefficients according to the method of Minguez-Mosquera et al. (1991). The extinction coefficients applied were E0 = 613 for pheophytin as the major component of the chlorophyll fraction, and E0 = 2000 for lutein as the major component of the carotenoid fraction. The chlorophyll and carotenoid content were expressed as milligrams of pheophytin or lutein per kilogram of oil, respectively, and were calculated (Haddada et al., 2007) as follows:

$$\text{Chlorophyll content (mg / kg)} = (A_{670} \times 10^6)/(613 \times 100 \times d)$$

$$\text{Carotenoid content (mg / kg)} = (A_{470} \times 10^6)/(2000 \times 100 \times d)$$

where A is the absorbance and d is the spectrophotometer cell thickness (1 cm).

Determination of Triglyceride Composition

The analysis of TGs was performed according to the official chromatographic method of the EC (EC Regulation 2568/91), using reversed-phase high performance liquid chromatography (RP-HPLC). The apparatus was Shimadzu (Model LC-10AD) HPLC system (Shimadzu Co., Kyoto, Japan) coupled with a refractive index detector (Shimadzu, RID-10A) and connected to a computer equipped with the analytical program Shimadzu Class-Vp (Chromatography Data System, version 4.0) for the processing of the data. The 5% solution of olive oil sample dissolved in acetone was injected into the HPLC system via Rheodyne injection valve with a 20 μl fixed loop (Rheodyne, CA, USA). The chromatographic separation was achieved on a Lichrospher 100 RP-18 analytical column (250 × 4 mm i.d.; 5μm particle size) at 40 °C, using a CTO-10A (Shimadzu) HPLC oven. A mixture of acetone/acetonitrile (60:40, v/v) was used as the elution solvent (mobile phase). Isocratic elution of TGs was carried out at a flow rate of 0.7 ml/min in 45 min. The TGs in olive oils were separated according to the equivalent carbon number (ECN), defined as CN – 2n, where CN is the carbon number of three fatty acids on the TG molecule and n is the total number of double bonds.

Statistical Analysis

The statistical data processing was performed using the SPSS 17.0 Statistics software (SPSS Inc.). The comparison of the means was achieved by one-way variance analysis (ANOVA). Linear Discriminant Analysis (LDA) was carried out to explore the possibility of classification olive oil samples according to cultivar and geographical origin.

The LDA is a multivariate probabilistic classification method based on the use of multivariate probability distribution with the hypothesis of normal distribution with the same variance-covariance matrix in all the considered classes. It must be emphasized that LDA can be used only with well-defined data matrixes that is those having high ratio between number of objects and number of variables inside each category (Boggia et al., 2002). In order to overcome the problem of the small number of olive oil samples per harvesting period, the two harvesting periods were considered as one with the assumption that olive oil samples originating from the same olive orchards under similar climatic conditions for the two harvesting periods would be similar in composition. The mathematical model built by applying the LDA procedure on the n selected variables consists of i linear classification functions (fi) of the form:

$$f_i = c_{i1}v_1 + c_{i2}v_2 + \dots + c_{in}v_n$$

where $v_1, \ldots, v_n$ are the values of each variable and $c_{i1}, \ldots, c_{in}$ are the classification coefficients assigned to the respective variables.

DISCUSSION AND RESULTS

Conventional Quality Indices

All olive oil samples had conventional parameter values within the range established by the EU (acidity $\leq 0.8\%$, $K_{232} \leq 2.5$, $K_{270} \leq 0.22$, $\Delta K \leq 0.01$) for the category "extra virgin olive oil" (EC Regulation 1989/2003).

Values for parameters (acidity, K_{232}, K_{270}, ΔK) obtained from olive oil samples for each region and cultivar are presented in Tables 2 and 3, respectively. The region of Kefalonia and the Thiaki cultivar showed the lowest mean values for acidity, K_{232} and K_{270}. The conventional quality indices of olive oil samples from the Koroneiki cultivar are shown in Table 4. Acidity varied from 0.26% (Lefkada and Kefalonia) to 0.44% (Messologi), K_{232} ranged of 1.29 (Lefkada) to 1.71 (Messologi), and K_{270} was between 0.10 (Lefkada and Kefalonia) and 0.13 (Messologi).

Chlorophyll and Carotenoid Content

Values for chlorophyll and carotenoid content of olive oil samples from each geographical region and cultivar are presented in Tables 2 and 3, respectively. Significant differences between regions and cultivars were observed in chlorophylls and carotenoids. The concentrations of these pigments depend on olive cultivar, olive fruit ripeness index, the soil and climatic conditions, and agronomic procedures (Kiritsakis et al., 2002).

Messologi had the maximum chlorophyll (4.26 mg/kg) and carotenoid (2.64 mg/kg) content, while the olive oils from Preveza presented the lowest pigment mean values 1.57 mg/kg and 0.96 mg/kg, respectively. In case of cultivar, chlorophylls ranged from 1.65 mg/kg (Asprolia) to 3.36 mg/kg (Ntopia of Zakynthos), and carotenoids ranged from 1.01 mg/kg (Asprolia) to 1.78 mg/kg (Thiaki).

Chlorophyll and carotenoid content for the Koroneiki olive oil samples are shown in Table 4. Both showed important differences between geographical origins of the Koroneiki cultivar. The Koroneiki olive oils from Messologi had the highest mean values, while the samples from the Kerkyra presented the lowest amounts of pigments. The values of chlorophylls and carotenoids are in accordance with the typical ranges reported by other authors (Haddada et al., 2007; Minguez-Mosquera et al., 1991;).

Triglyceride Composition

The chromatographic TG elution profiles of the virgin olive oil samples studied were all found to be qualitatively similar. The peaks correspond to the TGs or mixture of TGs, as some of these peaks are complex. Thirty one different TGs (LLL, OLLn, PoLL, PLLn, OLL, PoOL, OOLn, PoPoO, PLL, PLnO, PPoL, PPoPo, OOL, PoOO, POL, SLL, PPL, EeOO, OOO, SOL, POO, POP, PSL, GOO, SOO, OLA, POS, SLS, AOO, SOS, POA) were identified by comparison of their retention time to available standards and by calculating the ECN. The TGs are designated by letters corresponding to abbreviated names of fatty acids fixed to glycerol as follows: P, palmitic; Po, palmitoleic; Ee, eptadecenoic; S, stearic; O, oleic; L, linoleic; Ln, linolenic; A, arachidic; and G, gadoleic or eicosenoic. Data of TG composition were expressed as percentage of total TGs. It was assumed that the sum of the areas of the peaks corresponding to the TGs or TG mixtures was equal to 100%. Thus, the relative percentage of each TG was calculated. Also the values for ECN42, ECN44, ECN46, ECN48, ECN50, and ECN52 were calculated.

The TG composition (%) of olive oils for each region and cultivar is shown in Tables 2 and 3, respectively. In all samples analyzed, the main TG peaks (in order of decreasing concentration) were OOO, SOL + POO, OOL + POOO, POL + SLL, SOO + OLA, and POP + PSL. These accounted for more than 90% of the total area of peaks in the chromatogram. The major TG in all samples was OOO (39.84%) due to the high proportion of oleic acid in olive oils. Aranda et al. (2004) also reported high concentration of OOO (51.70%) for Spanish (Cornicabra) virgin olive oil samples. The second peak in order of quantitative contribution was SOL + POO TG mixture with an average content of 24.10% similar to that (20.80%) reported by Aranda et al. (2004). The value of trilinolein (LLL) did not exceed the maximum limit of 0.5% determined by the EC (EC Regulation 2568/91) as also reported by Stefanoudaki et al. (1997) for olive oil samples produced from the Koroneiki and Mastoides cultivars. The presence of a high triolein (OOO) level in an olive oil is inversely proportional to LLL and constitutes a favorable olive oil authenticity indicator (Ollivier et al., 2003).

As shown in Tables 2 and 3, the TG composition exhibited a great variability depending on olive cultivar, but also other factors such as soil characteristics and climatic conditions, which were not controlled in the present study. In case of region, OOO varied between 34.92% (Preveza) and 42.17% (Messologi); OOL + POOO between 11.80% (Messologi) and 15.88% (Preveza); POL + SLL between 5.68% (Messologi) and 9.31% (Preveza); PLL between 0.53% (Lefkada) and 0.80% (Kerkyra); and LLL between 0.06% (Kerkyra) and 0.12% (Zakynthos). In case of cultivar, OOO varied between 35.58% (Lianolia) and 43.09% (Thiaki); OOL + PoOO between 11.46% (Koroneiki) and 15.60% (Asprolia); POL + SLL between 6.02% (Thiaki) and 8.96% (Lianolia); OLL + POL between 1.36% (Koroneiki) and 2.70% (Asprolia); EeOO between 0.07% (Asprolia), and 0.26% (Thiaki); and LLL between 0.06% (Thiaki) and 0.16% (Ntopia of Zakynthos). Studies have shown that reduction in temperature of a given climate related or not to an increase in altitude, induces an increase in the unsaturation of fatty acids on TGs, mainly in linoleic acid proportions (Aguilera et al., 2005). This is generally explained by the fact that low temperatures promote the activity of acyl-desaturases (Los and Murata, 1998).

The important statistical differences ($p < 0.05$) were observed in TG content that is SOS + POA, PLnO + PPoL + PPoPo, PLLn, SOO + OLA, PLL, AOO, and OLLn + PoLL between regions. Similar differences in TG content that is PLnO + PPoL + PPoPo, AOO, PLL, SOO + OLA, EeOO, OOL + PoOO, and PLLn were found between cultivars. The TG data obtained were compared to the literature (Stefanoudaki et al., 1997). Although experimental conditions vary between studies, the main TGs (OOO, POO, LOO, PLO) are comparable to those of the most common Spanish, Italian, Greek and Portuguese olive oil samples.

Table 2. Mean values of the quality indices, pigment content and TG composition (%) of EVOO samples for each geographical region.

	Zakynthos		Kefalonia		Lefkada		Kerkyra		Preveza		Messologi	
Acidity (%)	0.34 ± 0.10	a	0.22 ± 0.11	a	0.40 ± 0.21	a	0.44 ± 0.13	a	0.41 ± 0.16	a	0.44 ± 0.13	a
K_{232}	1.36 ± 0.09	a	1.31 ± 0.06	a	1.38 ± 0.09	ab	1.63 ± 0.16	b	1.73 ± 0.22	b	1.71 ± 0.00	b
K_{270}	0.12 ± 0.01	ab	0.10 ± 0.01	a	0.11 ± 0.01	a	0.13 ± 0.03	ab	0.11 ± 0.00	a	0.13 ± 0.00	b
ΔK	–0.001 ± 0.003	a	–0.001 ± 0.002	a	–0.001 ± 0.002	a	0.000 ± 0.003	a	0.000 ± 0.003	a	-0.003 ± 0.003	a
Chlorophylls (mg / kg)	3.42 ± 0.72	bc	2.42 ± 0.73	b	1.83 ± 0.31	ab	1.60 ± 0.70	ab	1.57 ± 0.09	a	4.26 ± 0.19	c
Carotenoids (mg / kg)	1.57 ± 0.30	b	1.65 ± 0.38	b	1.10 ± 0.18	ab	1.01 ± 0.34	ab	0.96 ± 0.03	a	2.64 ± 0.10	c
LLL	0.122 ± 0.059	a	0.072 ± 0.018	a	0.084 ± 0.038	a	0.058 ± 0.032	a	0.077 ± 0.023	a	0.097 ± 0.025	a
OLLn + PoLL	0.266 ± 0.030	b	0.250 ± 0.046	ab	0.269 ± 0.055	abc	0.331 ± 0.069	bc	0.351 ± 0.044	c	0.219 ± 0.002	a
PLLn	0.190 ± 0.019	a	0.180 ± 0.024	a	0.216 ± 0.037	abc	0.255 ± 0.034	c	0.254 ± 0.020	c	0.215 ± 0.003	b
OLL + PoOL	1.787 ± 0.659	a	1.590 ± 0.396	a	2.060 ± 0.982	ab	1.738 ± 0.845	ab	2.832 ± 0.326	b	1.449 ± 0.039	a
OOLn + PoPoO	1.908 ± 0.140	ab	1.825 ± 0.218	ab	1.726 ± 0.086	a	2.029 ± 0.245	ab	2.091 ± 0.087	b	1.799 ± 0.013	a
PLL	0.661 ± 0.064	abc	0.658 ± 0.112	abc	0.529 ± 0.093	a	0.799 ± 0.092	c	0.723 ± 0.060	bc	0.661 ± 0.020	b
PLnO + PPoL + PPoPo	0.079 ± 0.007	abc	0.089 ± 0.010	bc	0.067 ± 0.006	a	0.116 ± 0.030	c	0.120 ± 0.018	c	0.082 ± 0.003	b
OOL + PoOO	12.518 ± 1.154	a	12.090 ± 1.035	a	13.744 ± 2.840	ab	12.508 ± 3.010	ab	15.879 ± 0.507	b	11.796 ± 0.558	a
POL + SLL	6.703 ± 1.148	a	6.760 ± 1.218	a	6.450 ± 1.469	a	7.237 ± 2.086	ab	9.308 ± 0.595	b	5.677 ± 0.333	a
PPL	0.660 ± 0.177	a	0.693 ± 0.217	a	0.594 ± 0.193	a	0.797 ± 0.319	ab	1.089 ± 0.076	b	0.477 ± 0.035	a
EeOO	0.106 ± 0.016	ab	0.179 ± 0.087	ab	0.104 ± 0.059	ab	0.145 ± 0.023	b	0.104 ± 0.006	a	0.137 ± 0.030	ab
OOO	39.457 ± 3.467	ab	40.220 ± 3.733	ab	41.529 ± 4.542	b	38.877 ± 4.099	ab	34.921 ± 1.866	a	42.175 ± 0.853	b
SOL + POO	24.585 ± 0.870	a	24.409 ± 1.348	a	22.648 ± 1.246	a	24.875 ± 1.637	a	23.238 ± 0.074	a	23.971 ± 0.178	a
POP + PSL	3.432 ± 0.483	ab	3.457 ± 0.609	ab	2.800 ± 0.213	a	3.864 ± 0.373	b	3.936 ± 0.185	b	3.024 ± 0.024	a
GOO	0.336 ± 0.069	ab	0.341 ± 0.057	ab	0.332 ± 0.134	ab	0.422 ± 0.126	b	0.283 ± 0.008	a	0.441 ± 0.018	b
SOO + OLA	4.888 ± 0.432	bc	4.916 ± 0.460	bc	4.801 ± 0.436	b	3.967 ± 0.976	ab	3.197 ± 0.133	a	5.402 ± 0.142	c

Table 2. *(Continued)*

	Zakynthos		Kefalonia		Lefkada		Kerkyra		Preveza		Messologi	
POS + SLS	1,254 ± 0.057	b	1,215 ± 0.147	b	1.127 ± 0.192	b	1.024 ± 0.219	ab	0.904 ± 0.002	a	1.247 ± 0.051	b
AOO	0.757 ± 0.098	bc	0.763 ± 0.083	bc	0.663 ± 0.111	b	0.647 ± 0.119	b	0.519 ± 0.006	a	0.817 ± 0.006	c
SOS + POA	0.290 ± 0.021	bc	0.291 ± 0.028	bc	0.257 ± 0.037	b	0.202 ± 0.034	ab	0.176 ± 0.002	a	0.317 ± 0.012	c
ECN42	0.579 ± 0.095	ab	0.503 ± 0.082	a	0.569 ± 0.127	ab	0.644 ± 0.130	ab	0.682 ± 0.088	b	0.530 ± 0.023	a
ECN44	4.436 ± 0.760	a	4.162 ± 0.552	a	4.382 ± 0.898	ab	4.682 ± 1.170	ab	5.765 ± 0.491	b	3.990 ± 0.003	a
ECN46	19.986 ± 2.449	a	19.722 ± 2.227	a	20.892 ± 4.425	ab	20.688 ± 5.347	ab	26.379 ± 1.172	b	18.087 ± 0.895	a
ECN48	67.475 ± 2.888	b	68.086 ± 2.575	b	66.976 ± 5.428	ab	67.616 ± 5.108	ab	62.095 ± 1.607	a	69.170 ± 1.055	b
ECN50	6.477 ± 0.494	bc	6.473 ± 0.479	bc	6.260 ± 0.543	b	5.413 ± 1.304	ab	4.383 ± 0.140	a	7.089 ± 0.175	c
ECN52	1.047 ± 0.115	bc	1.054 ± 0.100	bc	0.920 ± 0.122	b	0.849 ± 0.149	ab	0.696 ± 0.004	a	1.134 ± 0.005	c

Table 3. Mean values of the quality indices, pigment content and TG composition (%) of EVOO samples for each cultivar.

	Koroneiki		Ntopia of Zakynthos		Thiaki		Asprolia		Lianolia	
Acidity (%)	0.31 ± 0.12	ab	0.41 ± 0.08	b	0.18 ± 0.03	a	0.48 ± 0.21	b	0.44 ± 0.15	b
K_{232}	1.40 ± 0.16	abc	1.43 ± 0.08	b	1.25 ± 0.03	a	1.43 ± 0.05	ab	1.71 ± 0.15	c
K_{270}	0.11 ± 0.02	ab	0.12 ± 0.01	b	0.09 ± 0.01	a	0.11 ± 0.01	ab	0.13 ± 0.03	ab
ΔK	–0.001 ± 0.002	a	0.000 ± 0.002	a	–0.001 ± 0.002	a	0.000 ± 0.002	a	-0.001 ± 0.003	a
Chlorophylls (mg/kg)	2.92 ± 1.03	b	3.36 ± 0.98	b	1.91 ± 0.29	ab	1.65 ± 0.20	ab	1.78 ± 0.58	ab
Carotenoids (mg/kg)	1.58 ± 0.55	ab	1.51 ± 0.39	ab	1.78 ± 0.36	b	1.01 ± 0.14	a	1.09 ± 0.28	a
LLL	0.074 ± 0.034	ab	0.163 ± 0.056	b	0.064 ± 0.026	a	0.108 ± 0.015	b	0.078 ± 0.023	ab
OLLn + PoLL	0.256 ± 0.041	ab	0.275 ± 0.026	b	0.213 ± 0.028	a	0.303 ± 0.024	b	0.358 ± 0.058	b
PLLn	0.198 ± 0.026	ab	0.189 ± 0.017	a	0.161 ± 0.013	a	0.239 ± 0.016	b	0.264 ± 0.026	b
OLL + PoOL	1.357 ± 0.408	a	2.227 ± 0.735	ab	1.557 ± 0.467	ab	2.703 ± 0.217	b	2.533 ± 0.565	b
OOLn + PoPoO	1.865 ± 0.129	ab	1.963 ± 0.150	b	1.619 ± 0.144	a	1.726 ± 0.109	ab	2.146 ± 0.139	b
PLL	0.690 ± 0.056	b	0.653 ± 0.092	b	0.554 ± 0.078	ab	0.472 ± 0.037	a	0.797 ± 0.095	b
PLnO + PPoL + PPoPo	0.081 ± 0.008	b	0.083 ± 0.008	b	0.090 ± 0.015	b	0.066 ± 0.006	a	0.131 ± 0.016	c
OOL + PoOO	11.456 ± 1.262	a	13.186 ± 1.261	ab	12.198 ± 1.184	a	15.605 ± 0.508	b	15.133 ± 1.501	b
POL + SLL	6.095 ± 1.183	ab	7.515 ± 1.202	abc	6.016 ± 0.965	a	7.404 ± 0.411	b	8.963 ± 1.002	c
PPL	0.587 ± 0.193	a	0.786 ± 0.201	ab	0.538 ± 0.147	a	0.719 ± 0.059	a	1.054 ± 0.139	b
EeOO	0.130 ± 0.032	b	0.102 ± 0.003	b	0.259 ± 0.082	c	0.067 ± 0.014	a	0.123 ± 0.025	b
OOO	41.456 ± 3.482	ab	36.861 ± 3.081	a	43.092 ± 2.855	b	38.598 ± 1.336	a	35.580 ± 2.441	a
SOL + POO	24.671 ± 0.967	b	25.119 ± 0.597	b	23.230 ± 1.233	ab	21.956 ± 0.575	a	23.713 ± 1.344	ab
POP + PSL	3.357 ± 0.516	ab	3.832 ± 0.332	b	2.929 ± 0.360	a	2.826 ± 0.205	a	3.929 ± 0.379	b
GOO	0.408 ± 0.090	b	0.276 ± 0.061	ab	0.360 ± 0.052	b	0.246 ± 0.035	a	0.311 ± 0.033	ab
SOO + OLA	4.984 ± 0.411	b	4.543 ± 0.311	b	5.004 ± 0.712	b	4.901 ± 0.482	b	3.271 ± 0.314	a
POS + SLS	1.226 ± 0.147	bc	1.270 ± 0.038	c	1.065 ± 0.053	b	1.208 ± 0.149	bc	0.895 ± 0.075	a

Table 3. *(Continued)*

	Koroneiki		Ntopia of Zakynthos		Thiaki		Asprolia		Lianolia	
AOO	0.788 ± 0.054	b	0.679 ± 0.093	ab	0.778 ± 0.105	b	0.596 ± 0.061	a	0.543 ± 0.050	a
SOS + POA	0.290 ± 0.033	b	0.278 ± 0.016	b	0.273 ± 0.037	b	0.257 ± 0.038	b	0.178 ± 0.015	a
ECN42	0.527 ± 0.078	ab	0.627 ± 0.088	b	0.438 ± 0.059	a	0.650 ± 0.041	b	0.700 ± 0.104	b
ECN44	3.993 ± 0.525	a	4.926 ± 0.817	ab	3.819 ± 0.575	a	4.967 ± 0.243	b	5.607 ± 0.661	b
ECN46	18.268 ± 2.470	a	21.589 ± 2.641	ab	19.012 ± 2.241	a	23.795 ± 0.841	b	25.274 ± 2.497	b
ECN48	69.484 ± 2.967	b	65.811 ± 3.066	ab	69.251 ± 2.587	b	63.380 ± 0.717	a	63.222 ± 2.829	a
ECN50	6.619 ± 0.511	b	6.090 ± 0.367	b	6.430 ± 0.751	b	6.355 ± 0.599	b	4.477 ± 0.361	a
ECN52	1.078 ± 0.069	b	0.957 ± 0.108	b	1.051 ± 0.142	b	0.853 ± 0.084	ab	0.720 ± 0.057	a

The percentage of TG composition for the Koroneiki olive oil samples is shown in Table 4. The TG values showed considerable variability between regions (Zakynthos, Kefalonia, Lefkada, Kerkyra, and Messologi). For example, OOO varied between 38.17% (Kefalonia) and 47.39% (Lefkada); OOL + PoOO between 9.75% (Kerkyra) and 12.02% (Zakynthos); POL + SLL between 4.54% (Lefkada) and 7.29% (Kefalonia); OLL + PoOL between 0.77% (Lefkada) and 1.61% (Kefalonia); PPL between 0.34% (Lefkada) and 0.80% (Kefalonia); and LLL between 0.03% (Kerkyra) and 0.10% (Messologi). The mean values of the variables GOO, SOS + POA, PPL, OOO, POP + PSL, PLnO + PPoL + PPoPo, and POL + SLL, showed statistically significant differences ($p < 0.05$) between regions in the Koroneiki olive oil samples.

Classification of Olive Oils According to Cultivar

The differentiation and classification of the 48 olive oil samples into five cultivar groups (Koroneiki, Ntopia of Zakynthos, Thiaki, Asprolia, and Lianolia) was carried out by means of LDA. The LDA was applied to the data set (17 variables), obtained by the Fisher F ratio method (rejecting the variables with $p > 0.05$ and tolerance level $\leq$ 0.001). The discrimination of the five cultivars was based on TGs, with the following variables: LLL ($F = 9.5$), OLLn + PoLL ($F = 12.9$), PLLn ($F = 20.0$), OLL + PoOL ($F = 16.6$), OOLn + PoPoO ($F = 14.5$), PLL ($F = 22.7$), PLnO + PPoL + PPoPo ($F = 42.2$), OOL + PoOO ($F = 21.7$), POL + SLL ($F = 11.4$), PPL ($F = 11.5$), EeOO ($F = 22.2$), OOO ($F = 8.1$), SOL + POO ($F = 11.9$), POP + PSL ($F = 8.1$), GOO ($F = 9.0$), POS + SLS ($F = 11.8$), SOS + POA ($F = 18.9$).

The leave-one-out validation method was used to test the prediction classification ability. The presentation of the scores for each sample on the plane of the two canonical discriminant functions is shown in Figure 1(a). The correct classification rate, obtained with the leave-one-out method of cross-validation, was 89.6%. Thus, 43 of the olive oil samples were correctly classified in their cultivar group, while five samples were classified in another group. More specifically, three samples of Koroneiki were classified as Ntopia of Zakynthos, one sample of Ntopia of Zakynthos was classified as Koroneiki, and one sample of Thiaki were classified as Koroneiki. All olive oil samples belonging to cultivars Asprolia and Lianolia were correctly classified (100.0%). The discrimination between Koroneiki (87.5%), Ntopia of Zakynthos (83.3%), and Thiaki (80.0%) was not complete, even though these groups had a satisfactory classification rate, higher than 80.0%.

The data of TG composition and pigment content (chlorophylls and carotenoids) were combined and resubmitted to LDA. Nineteen variables were selected by the Fisher F ratio method, for the classification of the 48 olive oil samples into five cultivar groups (Koroneiki, Ntopia of Zakynthos, Thiaki, Asprolia, Lianolia). The selected variables were the following: LLL ($F = 9.5$), OLLn + PoLL ($F = 12.9$), PLLn ($F = 20.0$), OLL + PoOL ($F = 16.6$), OOLn + PoPoO ($F = 14.5$), PLL ($F = 22.7$), PLnO + PPoL + PPoPo ($F = 42.2$), OOL + PoOO ($F = 21.7$), POL + SLL ($F = 11.4$), PPL ($F = 11.5$), EeOO ($F = 22.2$), OOO ($F = 8.1$), SOL + POO ($F = 11.9$), POP + PSL ($F = 8.1$), GOO ($F = 9.0$), POS + SLS ($F = 11.8$), SOS + POA ($F = 18.9$), Chlorophylls ($F = 6.1$), Carotenoids ($F = 3.7$).

Table 4. Mean values of the quality indices, pigment content and TG composition (%) of the Koroneiki EVOO samples.

	Zakynthos		Kefalonia		Lefkada		Kerkyra		Messologi	
Acidity (%)	0.30 ± 0.09	a	0.26 ± 0.13	a	0.26 ± 0.14	a	0.40 ± 0.09	a	0.44 ± 0.13	a
K_{232}	1.31 ± 0.06	a	1.35 ± 0.05	a	1.29 ± 0.06	a	1.55 ± 0.20	ab	1.71 ± 0.00	b
K270	0.12 ± 0.01	ab	0.10 ± 0.01	a	0.10 ± 0.01	a	0.12 ± 0.03	ab	0.13 ± 0.00	b
ΔK	–0.001 ± 0.003	ab	–0.001 ± 0.002	ab	–0.003 ± 0.000	a	0.002 ± 0.002	b	-0.003 ± 0.003	ab
Chlorophylls (mg kg – 1)	3.47 ± 0.52	c	2.79 ± 0.74	bc	2.18 ± 0.07	b	1.13 ± 0.00	a	4.26 ± 0.19	d
Carotenoids (mg kg – 1)	1.61 ± 0.24	b	1.56 ± 0.39	b	1.28 ± 0.14	b	0.77 ± 0.00	a	2.64 ± 0.10	c
LLL	0.091 ± 0.041	b	0.078 ± 0.008	b	0.037 ± 0.005	a	0.031 ± 0.004	a	0.097 ± 0.025	b
OLLn + PoLL	0.260 ± 0.033	c	0.277 ± 0.036	c	0.201 ± 0.011	a	0.288 ± 0.035	c	0.219 ± 0.002	b
PLLn	0.192 ± 0.022	ab	0.194 ± 0.021	ab	0.170 ± 0.007	a	0.234 ± 0.028	b	0.215 ± 0.003	b
OLL + PoOL	1.457 ± 0.362	b	1.614 ± 0.375	b	0.775 ± 0.137	a	0.978 ± 0.075	a	1.449 ± 0.039	b
OOLn + PoPoO	1.866 ± 0.125	bc	1.973 ± 0.113	c	1.726 ± 0.002	a	1.817 ± 0.140	abc	1.799 ± 0.013	b
PLL	0.667 ± 0.040	a	0.732 ± 0.058	a	0.643 ± 0.048	a	0.728 ± 0.056	a	0.661 ± 0.020	a
PLnO + PPoL + PPoPo	0.077 ± 0.004	ab	0.088 ± 0.006	b	0.070 ± 0.005	a	0.085 ± 0.007	b	0.082 ± 0.003	b
OOL + PoOO	12.017 ± 0.817	a	12.012 ± 1.006	a	10.021 ± 0.665	a	9.754 ± 1.492	a	11.796 ± 0.558	a
POL + SLL	6.093 ± 0.652	bc	7.292 ± 1.145	c	4.541 ± 0.108	a	5.280 ± 0.807	ab	5.677 ± 0.333	b
PPL	0.565 ± 0.076	bc	0.804 ± 0.194	c	0.344 ± 0.012	a	0.489 ± 0.075	b	0.477 ± 0.035	b
EeOO	0.108 ± 0.021	a	0.122 ± 0.016	ab	0.178 ± 0.033	b	0.155 ± 0.024	b	0.137 ± 0.030	ab
OOO	41.405 ± 2.332	ab	38.169 ± 2.886	a	47.390 ± 0.874	c	42.614 ± 0.701	b	42.175 ± 0.853	b
SOL + POO	24.185 ± 0.851	ab	25.251 ± 0.576	b	24.032 ± 1.036	ab	25.950 ± 0.429	b	23.971 ± 0.178	a
POP + PSL	3.133 ± 0.338	ab	3.834 ± 0.442	b	2.747 ± 0.264	a	3.784 ± 0.062	b	3.024 ± 0.024	a
GOO	0.380 ± 0.029	a	0.328 ± 0.060	a	0.506 ± 0.032	c	0.541 ± 0.093	bc	0.441 ± 0.018	b
SOO + OLA	5.146 ± 0.316	ab	4.854 ± 0.199	a	4.602 ± 0.303	a	4.822 ± 0.825	ab	5.402 ± 0.142	b
POS + SLS	1.242 ± 0.068	b	1.322 ± 0.075	b	0.965 ± 0.181	a	1.204 ± 0.206	ab	1.247 ± 0.051	b

Table 4. *(Continued)*

	Koroneiki		Ntopia of Zakynthos		Thiaki		Asprolia		Lianolia	
AOO	0.816 ± 0.048	a	0.753 ± 0.070	a	0.797 ± 0.001	a	0.763 ± 0.033	a	0.817 ± 0.006	a
SOS + POA	0.300 ± 0.019	b	0.304 ± 0.010	b	0.258 ± 0.044	ab	0.234 ± 0.010	a	0.317 ± 0.012	b
ECN42	0.543 ± 0.089	b	0.549 ± 0.062	b	0.407 ± 0.022	a	0.553 ± 0.067	b	0.530 ± 0.023	b
ECN44	4.068 ± 0.483	bc	4.407 ± 0.414	c	3.213 ± 0.082	a	3.608 ± 0.278	b	3.990 ± 0.003	c
ECN46	18.784 ± 1.516	b	20.230 ± 2.241	b	15.085 ± 0.818	a	15.678 ± 2.398	ab	18.087 ± 0.895	b
ECN48	68.722 ± 2.151	a	67.254 ± 2.399	a	74.169 ± 0.427	c	72.348 ± 1.192	b	69.170 ± 1.055	a
ECN50	6.768 ± 0.363	ab	6.504 ± 0.207	a	6.072 ± 0.452	a	6.567 ± 1.123	ab	7.089 ± 0.175	b
ECN52	1.116 ± 0.062	bc	1.057 ± 0.070	ab	1.054 ± 0.043	b	0.997 ± 0.043	a	1.134 ± 0.005	c

The class-separation is presented in Figure 1(b). Analysis showed that the increase of the correct classification rate of the olive oil samples, given by the leave-one-out method of cross-validation, was small. Forty four of the olive oil samples (91.7%) were correctly classified in the group corresponding to their cultivar, while four samples were classified in another group. More specifically, three samples of Koroneiki were classified as Ntopia of Zakynthos, and one sample of Ntopia of Zakynthos was classified as Koroneiki. The correct classification rate of the cultivar groups Koroneiki (87.5%), Ntopia of Zakynthos (83.3%), Asprolia (100.0%), and Lianolia (100.0%) remained unchanged. On the contrary, the correct prediction rate for the cultivar Thiaki (100.0%) was satisfactorily increased. Classification rates lower than 100% may be related to similarities of botanical origin of each cultivar (Stefanoudaki et al., 1997).

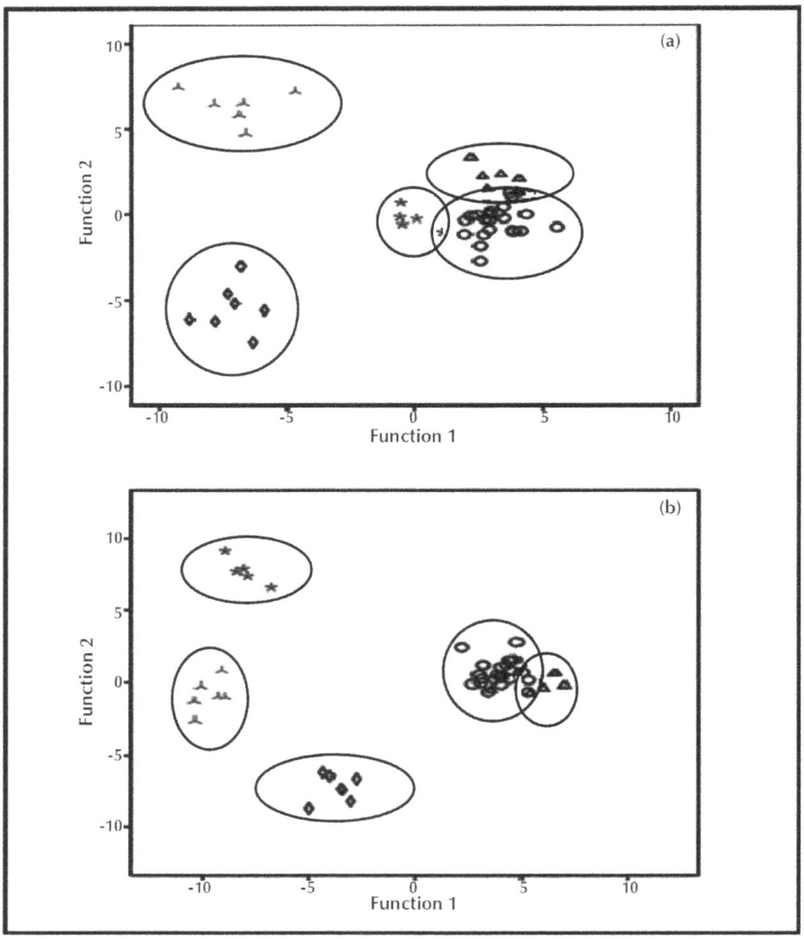

Figure 1. Plane representation of discriminant functions for the discrimination of the olive oil samples according to cultivar, based on (a) the TG composition (b) the TG composition and pigment content. (O) Koroneiki, (Δ) Ntopia of Zakynthos, (✳) Thiaki, (⅄) Asprolia, (◇) Lianolia.

Ollivier et al (2003) analyzed 564 French olive oil samples for fatty acid and TG composition and squalene from 6 different cultivars. The stepwise LDA of TG data led to the selection of 20 variables based on which correct classification reached 100%. As such perfect classification was attributed to the absolutely certain identification of the cultivars selected for the analysis. Diaz et al. (2005) analyzed 80 Spanish EVOOs for TGs and sterols and applied PCA and soft independent modeling class analogy (SIMCA) to differentiate samples with regard to cultivar. They reported that the PPP, SOO, and OOO TGs were the most important variables for the differentiation of cultivars of olive oils. Of the sterols analyzed β-sitosterol, Δ-5-avenasterol, total β-sitosterol and campesterol were the most important variables to characterize olive oils according to cultivar. Aranda et al. (2004) analyzed 224 commercial Spanish olive oil samples for TG, total and 2-position fatty acid composition belonging to four different varieties using both PCA and DA. The main TG peaks identified were: OOO, SOL + POO, OLO + LnPP and OLA + SOO accounting for more than 85% of the total peak area in the chromatogram. Based on these four variables a reasonable percentage of correct classification (86–99%) of each of the virgin olive oil varieties studied was achieved. Statistical analysis suggested that TG variables were more important than total and 2-position FA for optimum classification of commercial samples analyzed. Finally, Bronzini de Caraffa et al (2008) analyzed 89 samples of Corsican virgin olive oils belonging to nine different cultivars for TG and fatty acid composition. Using the random-amplified polymorphic DNA technique, they showed that the nine different cultivars belonged to four basic varieties. Based on stepwise factorial discriminant (FDA) analysis they reported that discrimination of the nine cultivars was poor. However respective correct classification of samples in the corresponding four varieties was 100%. The Discriminating variables included: OOO (F = 131.6), OOL (F = 104.5), PoOO (F = 76.1), 18:0 (F = 62.1) and OLL (F = 25.9).

Classification of Koroneiki Olive Oils According to Geographical Origin

The LDA was performed for the geographical classification of the olive oil samples belonging to Koroneiki cultivar (24 samples) into five groups (Zakynthos, Kefalonia, Lefkada, Kerkyra, and Messologi) based on TG composition. The application of discriminant analysis led to the selection of variables by the Fisher F ratio method to determine discriminant functions (rejecting the variables with $p > 0.05$ and tolerance level ≤ 0.001). The 13 selected variables were the following: LLL (F = 5.0), OLLn + PoLL (F = 5.1), PLLn (F = 4.7), OLL + PoOL (F = 5.3), OOLn + PoPoO (F = 3.3), PLL (F = 3.4), PLnO + PPoL + PPoPo (F = 7.9), OOL + PoOO (F = 5.7), POL + SLL (F = 7.6), PPL (F = 9.8), EeOO (F = 6.5), SOL + POO (F = 5.8), POP + PSL (F = 8.9).

The leave-one-out validation method was used to test the prediction classification ability. The presentation of the scores for each sample on the plane of the two canonical discriminant functions is shown in Figure 2. The correct classification rate, obtained with the leave-one-out method of cross-validation, was 83.3%. Thus, 20 of the Koroneiki olive oil samples were correctly classified in their geographical origin group, while four samples were classified in another group. More specifically, two samples of Zakynthos were classified in Kefalonia, and two samples of Kefalonia were classified in Zakynthos (1 sample) and in Lefkada (1 sample). All olive oil samples belonging to regions Lefkada, Kerkyra, and Messologi were correctly classified (100.0%). The classification rate of samples from Zakynthos and Kefalonia was 75.0% and 71.4%, respectively.

The LDA was subsequently performed for the differentiation of the Koroneiki olive oils based on both TG composition and pigment content (chlorophylls and carotenoids). The classification results remained unchanged, because the variables chlorophylls and carotenoids were rejected by the Fisher F ratio method.

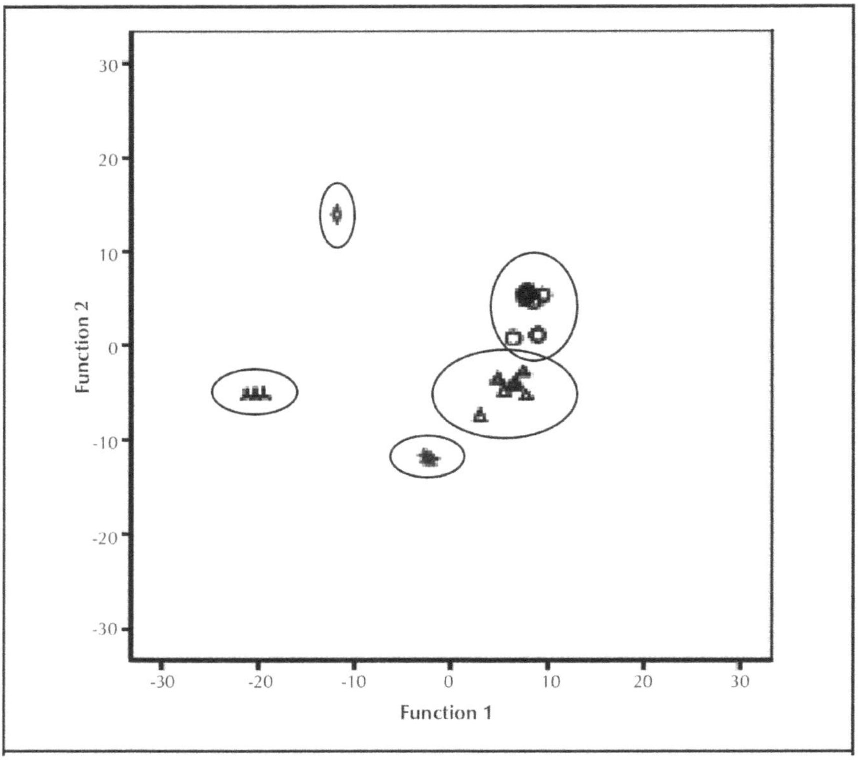

Figure 2. Plane representation of discriminant functions for the discrimination of the Koroneiki olive oil samples according to geographical origin. (O) Zakynthos, (Δ) Kefalonia, (✳) Lefkada, (◊) Kerkyra, (⅄) Lianolia.

Stefanoudaki et al (1997) analyzed the TG profile of 120 virgin olive oil samples collected at different maturity stages from the region of Chania, Greece belonging to the Koroneiki and Mastoides cultivars in an attempt to classify Cretan olive oils according to geographical origin. The TG data were coupled to PCA and clustering analysis (CA). Based on 20 different TGs determined, the oils were grouped quite clearly according to their geographical origin within each cultivar. Petrakis et al (2008) analyzed 131 EVOOS (cv. Koroneiki) from six sites belonging to three different regions of Southern Greece, by means of 1H and 31P NMR spectroscopy and classified them according to their content in fatty acids, phenolics, diacylglycerols, total free sterols, free acidity, and iodine number. Discriminant analysis was applied for the geographical prediction of Koroneiki olive oil samples. The results showed that the correct classification rate at the level of three regions was 87.0%, while at the level of six sites was 74.0%.

Classification of Olive Oils According To Geographical Origin

In this case, LDA was performed to classify the olive oil samples from all cultivars according to geographical origin into six groups (Zakynthos, Kefalonia, Lefkada, Kerkyra, Preveza, and Messologi). The application of LDA to TG composition data led to the selection of variables by the Fisher F ratio method to determine discriminant functions (rejecting the variables with $p > 0.05$ and tolerance level ≤ 0.001). The 14 selected variables were the following: LLL ($F = 3.2$), OLLn + PoLL ($F = 5.3$), PLLn ($F = 10.1$), OOLn + PoPoO ($F = 3.9$), PLL ($F = 7.9$), PLnO + PPoL + PPoPo ($F = 14.2$), OOL + PoOO ($F = 2.7$), POL + SLL ($F = 2.7$), PPL ($F = 3.6$), EeOO ($F = 3.3$), SOL + POO ($F = 4.3$), POP + PSL ($F = 6.1$), AOO ($F = 5.6$), SOS + POA ($F = 19.8$).

The leave-one-out validation method was used to test the prediction classification ability. The class-separation is presented in Figure 3(a), which shows the scores of each sample on the plane of the two canonical discriminant functions. The correct classification rate, obtained with the leave-one-out method of cross-validation, was 75.0%. Thus, 36 of the olive oil samples were correctly classified in their geographical origin group, while 12 samples were classified in another group. More specifically, two samples of Zakynthos were classified in Kefalonia (1 sample) and in Messologi (1 sample); five samples of Kefalonia were classified in Zakynthos (3 samples), in Lefkada (1 sample) and in Messologi (1 sample); two samples of Lefkada were classified in Zakynthos (1 sample) and in Kefalonia (1 sample); and three samples of Kerkyra were classified in Preveza. All olive oil samples from Preveza and Messologi were correctly classified (100.0%). The origin groups Zakynthos (85.7%) and Lefkada (77.8%) had a high classification rate. On the contrary, the prediction classification ability for the groups Kefalonia (58.3%) and Kerkyra (57.1%) was less satisfactory. The Kefalonia and Kerkyra samples were not distinctly separated from the Zakynthos and Preveza samples respectively, most probably because of the similar climatic conditions of these regions and of the same cultivar (Koroneiki and Lianolia, respectively) of these samples. It is well-documented that TG content of olive oils is strongly dependent on the cultivar (Bronzini de Caraffa et al., 2008; Fontanazza et al., 1994; Stefanoudaki et al., 1997).

To further increase the classification ability of the LDA method, a new discriminant analysis based on both TG composition and pigment content (chlorophylls and carotenoids) was performed. For the differentiation and classification of the 48 olive oil samples into six origin groups (Zakynthos, Kefalonia, Lefkada, Kerkyra, Preveza, and, Messologi), sixteen variables were selected by the Fisher F ratio method. The 16 selected variables were the following: LLL ($F = 3.2$), OLLn + PoLL ($F = 5.3$), PLLn ($F = 10.1$), OOLn + PoPoO ($F = 3.9$), PLL ($F = 7.9$), PLnO + PPoL + PPoPo ($F = 14.2$), OOL + PoOO ($F = 2.7$), POL + SLL ($F = 2.7$), PPL ($F = 3.6$), EeOO ($F = 3.3$), SOL + POO ($F = 4.3$), POP + PSL ($F = 6.1$), AOO ($F = 5.6$), SOS + POA ($F = 19.8$), Chlorophylls ($F = 17.3$), Carotenoids ($F = 18.3$).

The class-separation is presented in Figure 3(b). Analysis showed that the correct classification rate of the samples, given by the leave-one-out method of cross-validation increased. Thus, 40 of the olive oil samples (83.3%) were correctly classified in the group corresponding to their geographical origin, while eight samples were classified in another group. More specifically, two samples of Zakynthos were classified in Kefalonia (1 sample) and in Messologi (1 sample); two samples of Kefalonia were classified in Messologi; one sample of Lefkada was classified in Zakynthos (1 sample); and three samples of Kerkyra were classified in Preveza. The correct classification rate of olive oil samples from Kefalonia (83.3%) and Lefkada (88.9%) was satisfactorily increased, while the rate of the samples belonging to origin groups Preveza (100.0%), Messologi (100.0%), Zakynthos (85.7%), and Kerkyra (57.1%) remained unchanged. Tsimidou et al (1987) used TG and fatty acid composition along with principal component analysis (PCA) to classify 42 Greek virgin olive oil samples of different cultivars collected from various locations over two harvesting seasons and demonstrated a higher level of discrimination regarding geographical origin, cultivar and time of harvesting using TG versus fatty acid analysis.

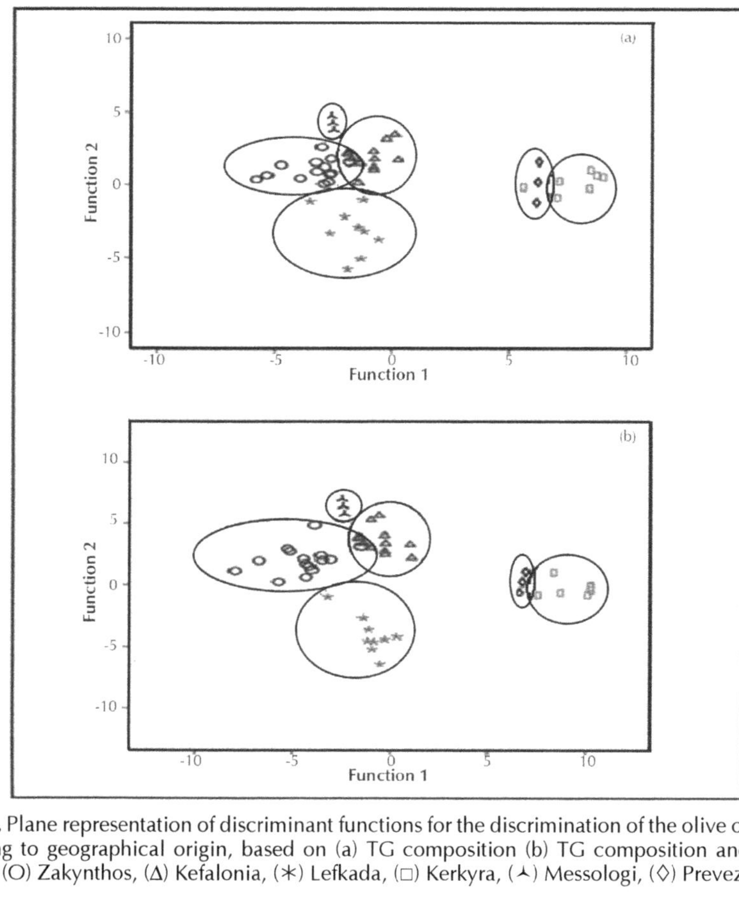

Figure 3. Plane representation of discriminant functions for the discrimination of the olive oil samples according to geographical origin, based on (a) TG composition (b) TG composition and pigment content. (O) Zakynthos, (Δ) Kefalonia, (∗) Lefkada, (□) Kerkyra, (⅄) Messologi, (◊) Preveza.

CONCLUSION

In the present study, LDA was successfully used to classify Greek EVOOs according to cultivar and geographical origin using TG analysis and pigment content. Free acidity, specific absorption coefficients (K_{232}, K_{270}), pigments (chlorophylls and carotenoids) and TG composition were determined in each of 48 samples. Results obtained showed significant statistical differences ($p < 0.05$) between regions of olive oil production and cultivars. Using the TG composition data, the olive oil samples were sufficiently classified according to cultivar and geographical origin. Combination of TG composition data with pigment content data increased the classification rate of olive oil samples.

Therefore, the combination of relatively simple to determine analytical parameters such as TGs and pigments (chlorophylls and carotenoids) with chemometric techniques such as LDA represents a useful tool to characterize and classify Western Greek extra virgin olive oils.

KEYWORDS

- **Cultivar**
- **Extra virgin olive oils**
- **Geographical origin**
- **Greek olive oils**
- **Linear discriminant analysis**
- **Olive oil**

ACKNOWLEDGMENTS

This study has been carried out with financial support from project LOC.elaion No. I2101021 within the INTERREG III GREECE-ITALY program.

Chapter 4

Effect of Irradiation on the Migration of ATBC Plasticizer from P (VDC-VC) Films into Fish Fillets

Panagiota D. Zygoura, Kyriakos A. Riganakos, and Michael G. Kontominas

INTRODUCTION

Cod and herring fillets were wrapped in plasticized polyvinylidene chloride/polyvinyl chloride (PVDC/PVC) film ("saran" wrap) containing acetyl tributyl citrate (ATBC) as plasticizer. The ratio of film surface to weight of food was ca. 4 dm^2 to 45 g fish (89:1). Wrapped samples of fish were subjected to high energy electron beam irradiation at doses equal to 5 and 10 kGy. Samples were subsequently stored at 4°C and analyzed for ATBC content at time intervals between 12 and 240 hr of contact (kinetic study). The determination of ATBC was performed by applying a simple, rapid and convenient extraction method followed by analysis on a GC apparatus coupled with a FI detector. The equilibrium migration concentrations of ATBC plasticizer in cod fillets ranged from 11.1 to 12.8 mg/kg, while the corresponding values for herring samples were between 32.4 and 33.4 mg/kg. Migration data showed that irradiation with high energy electrons at pasteurizing doses did not considerably affect the film's specific migration characteristics. Statistically significant differences ($p < 0.05$) were only observed between non-irradiated and irradiated at 10 kGy cod fillet samples. On the contrary, statistically significant differences in plasticizer migration were found between the two fish species examined. Thus, fat content of the packaged fish fillets substantially affected the extent to which migration of ATBC occurred. Plasticizer concentrations in both marine species exceeded the 5 mg/kg restriction (EU Synoptic Document, 2005). Non-compliance of the saran film used with EU specifications is due to the particular experimental design of the present study. However, household food overwrapping or rewrapping is often carried out under similar as above conditions.

Ionizing radiation, including gamma and electron beam radiation, is being used at various dose levels for food preservation as well as sterilization of packaging materials and medical products (Loaharanu, 2003; Molins, 2001; Murano, 1995; Wilkinson and Gould, 1996). The Joint FAO/IAEA/WHO Expert Committee on the Wholesomeness of Irradiated Food (JECFI) held in Geneva (1980) concluded that irradiation at a dose up to 10 kGy does not impart any toxicological risk to the food and does not affect its nutritional value (FICDB, 2010).

Fish comprises an ideal substrate for the growth of spoilage microorganisms (e.g. genera including *Pseudomonas*, *Bacillus*, *Shewanella*, *Psychrobacter*, *Flavobacterium*) and pathogens (including *Vibrio* spp., *Aeromonas hydrophila*, *Clostridium botulinum*, *Salmonella* spp., *Listeria monocytogenes*). Irradiation of fish and shellfish at low (< 1 kGy) and medium (1–10 kGy) doses of ionizing radiation combined with refrigeration and proper packaging, significantly prolongs the shelf life of such products through a drastic reduction of their microbial load. In addition, public health is ensured via elimination of pathogens (Andrews et al., 1995; Cook, 2001; Komolprasert and Morehouse, 2004; Molins, 2001; Palumbo et al,. 1986; Poole et al., 1994; Thibault and Charbonneau, 1991).

Products such as fish fillets are usually packaged in plastic films before being subjected to irradiation in order to avoid microbial recontamination. Thus, the effect of radiation on plastic packaging materials is of prime importance as it relates directly to the quality and/or safety of packaged foodstuffs. Depending on the particular material, nature of the additives used to compound the plastic, processing history and specific conditions of irradiation (absorbed dose, dose rate, temperature, and atmosphere), changes in the polymer structure may occur affecting, among others, its migration characteristics. According to the literature, irradiation may result in: (a) formation of cross-links and (b) scission of polymer chains. The first phenomenon is expected to repress migration of additives, while the second one is expected to enhance it (Buchalla et al., 1993a, 1993b, and 2000; Deschenes et al., 1995; Gheysari et al., 2001; Goulas et al., 2004; Sears and Darby, 1982; Zygoura et al., 2007).

Among potential migrants, plasticizers have raised considerable concern within the scientific community from the food safety point of view, since they are present in plastics films in significant amounts (up to 30% w/w in PVC). The ATBC (at levels of up to 5% w/w) is the most widely used plasticizer in vinylidene chloride films copolymerized with up to 20% vinyl chloride (PVDC/PVC). This type of film is being widely used for the wrapping of cheese, frozen meat/poultry and bakery products at a retail level and is also available as a domestic film wrap (Brody and Marsh, 1997; Castle et al., 1988a). With regard to the toxicity of plasticizers, the Scientific Committee on Food (SCF) has placed ATBC on their list 7 of 1995, under *Substances for which some toxicological data exist, but for which an ADI or a tolerable daily intake (TDI) could not be established.* In this case, the EU Synoptic Document (2005) lists a restriction of 5 mg/kg food.

There are numerous studies published in the literature with respect to ATBC migration. Several methods, based on either measuring the loss of citrate from the film using infrared spectroscopy, or determining (GC/FID-MS) the migrant after proper extraction from the matrix and separation from the co- extracted matter, have been employed to assess migration from saran films into food systems (e.g. meat, poultry and cheese) especially during microwave cooking (Badeka et al., 1999; Castle et al., 1988a, 1988b; Heath and Reilly, 1981; Sendon-Garcia et al., 2006; Van Lierop and Van Veen, 1988). However, to the best of our knowledge, there are no published data on the migration of plasticizers from flexible films into fish as affected by irradiation.

Thus, the objectives of the present study were (a) to propose a simple, rapid and efficient analytical method for the determination of plasticizers in complex food matrices, (b) to carry out a kinetic study on the migration of ATBC from saran film into marine fish species based on the aforementioned approach, and (c) to check compliance of the selected packaging material with EU restrictions when irradiation is combined with food overwrapping.

MATERIALS AND METHODS

Materials and Reagents

Plasticized PVDC/PVC film was commercial saran wrap (12 μm in thickness) supplied by Saropack A. G (Rorschach, Switzerland). The level of ATBC plasticizer was disclosed by the manufacturer as being (4.5 ± 0.1) % w/w, which was confirmed by chloroform extraction of the film followed by capillary GC analysis.

Analytical grade ATBC plasticizer was purchased from Unitex Chemical (N.C, USA). Analytical grade octadecane (≥99.0%), used as internal standard (IS), was purchased from Fluka (Buchs, Switzerland). Analytical grade *n*-hexane and anhydrous sodium sulphate (≥99.0%) were purchased from Merck (Darmstadt, Germany).

Fish Samples

Fresh cod and herring (*Gadus morhua* and *Argentina silus*) were purchased from retail outlets located in Karlsruhe (Germany), where it had been established that they had not previously been in contact with plasticized film. Blank measurements were carried out alongside migration experiments to examine whether fish samples were ATBC contaminated. The specific fish species were selected due to substantial differences in fat content.

Lipid and Moisture Analysis

Fat content was determined by the AOAC Soxhlet method (AOAC, 1995). Moisture content was determined gravimetrically; 5 g of minced fish fillet was dried at 105°C until constant weight.

Irradiation and Migration Experiments

Cod and herring were manually filleted. Fillets (with skin) of approximately the same weight and surface area were brought into contact (wrapped) with saran film. The average weight of fish samples was 45 ± 5 g and the total film/fish contact area was ca. 400 cm^2 for each sample. The ratio of film surface to weight of food was ca. 4 dm^2 to 45 g fish (89:1) in contrast to the generally agreed relationship of 6 dm^2 to 1 kg food (6:1) (EEC, 1990). The former ratio corresponds to more realistic food packaging applications.

Wrapped fish samples were irradiated using a linear accelerator (LINAC - CIRCE III, Linac Technologies S. A.; Orsay, France, 10 MeV) at doses equal to 5 and 10 kGy. Absorbed doses were measured using an Alanine - ESR dosimeter and mean dose rate was 10^7 Gy/s for all samples. Irradiation was carried out at 0 to (4 ± 2)°C, using ice, and samples were subsequently stored at 4 ± 1°C. Irradiation was conducted at the Federal Research Center for Nutrition and Food (Karlsruhe, Germany).

Sampling for plasticizer determination was carried out at predetermined time intervals; namely 12, 24, 48, 96, 144, 192, and 240 hr. By the end of storage period, sensory characteristics of both marine fish species had begun to deteriorate; especially those with a higher fat content. Nevertheless, the aforementioned conditions were chosen on a strict experimental basis in order to (a) allow feasibility of a kinetic study and (b) examine the extent to which radiation dose affects the levels of plasticizer migration.

Identical non-irradiated (control) samples were also analyzed for ATBC content for comparison purposes.

Plasticizer Analysis

Sample Preparation

The ATBC plasticizer was determined by the method developed by Nerin et al. (1992) for DEHA determination in cheese after appropriate adaptation: Whole fish fillet samples (45 ± 5g) were extracted with *n*-hexane (75 ml) by applying ultrasonication (30°C for 15 min). An ultrasonic thermostated bath (Bandelin Sonorex RK 100, Germany) was used. The organic extract was decanted and samples were re extracted (twice) with fresh solvent (150 ml). Following the extraction step, fish samples were washed with ca. 25 ml of solvent. The extracts, including the portion used for washing, were combined and dried with the addition of anhydrous sodium sulphate (ca. 30 g). The mixture was then filtered to remove Na_2SO_4 and the filtrate was subsequently evaporated to dryness with the aid of a flash evaporator. The extracted plasticizer was finally collected with 10 ml of *n*-hexane to which one milliliter of an octadecane solution (0.1 mg/ml in *n*-hexane) had been added as an internal standard. All glassware used were heated at 200°C for 2 hr, cooled and then rinsed with *n*-hexane before use.

The fat interference was minimized because the sample is extracted rather than dissolved using hexane as extraction solvent. Moreover, the extraction process is carried out using an ultrasonic bath in which the plasticizer is quantitatively extracted without dissolving the fat.

The GC Equipment and Conditions of Analysis

For qualitative and quantitative determination of the selected analyte, a gas chromatograph HP 5890 series II (Hewlett-Packard, Wilmington, USA) with a flame ionization detector (FID) was used, equipped with a 30 m x 0.32 mm fused silica capillary column, coated with 0.25 μm film (HP-5, J. & W. Scientific, Folsom, USA). The carrier gas was helium at 75 kPa (2 ml/min).

The temperature of the injector was maintained at 250°C, while the detector was set at 320°C. The temperature program of the GC oven was the following: 200°C, held 1 min; rating 8°C/min to 280°C (3 min hold); rating 20°C/min to a final temperature of 300°C with a 15 min hold. Sample injections were alternated with pure solvent (*n*-hexane) injections to remove excess of co-extracted lipophilic compounds from the column. The absence of additional peaks in the area of interest proved that there was no interference of the fat matter. Injection volume was 1 μl and detection was performed at a split (1:5) mode.

Recovery Tests

Fillet samples from both fish species which had not been in contact with the packaging material were spiked with ATBC plasticizer. They were subsequently stored for a short period of time (e.g. overnight) and subjected to the extraction procedure described under Section 2.5.

Spiking was carried out by addition of various amounts of plasticizer from 1% stock solution in hexane at concentrations in the proximity of real sample extracts (5.2, 13.0, and 35.1 mg/kg).

Standard Curve

An appropriate amount (100 μl) of ATBC was diluted with *n*-hexane to prepare 1% (10,500 ppm) stock solution. Appropriate amounts of this solution were added to ca. 45 g fish in order to cover the range 2.6–78.0 mg/kg. The extraction procedure was applied and standard curve equation was obtained by linear regression.

Statistical Analysis

Experiments were replicated twice on different occasions. All analyses were run in triplicate for each replicate ($n = 6$). Average values and standard deviations were calculated for all data. Differences between pairs were defined by Student's t test and were considered to be significant at the $p \leq 0.05$ levels.

DISCUSSION AND RESULTS

Method Optimization

The quantitative determination of plastics additives in food matrices is associated with two main difficulties; namely, the low detection limit required and the diversity of potential interferences present in foodstuffs.

The first part of this study involved the determination of the parameters affecting the extraction process and hence requiring optimization in order to obtain maximum recoveries. Thus, an experimental design was set forth in order to optimize and evaluate analytical characteristics of the method.

Effect of Extraction Solvent and Method

Alkanes, as well as their mixtures with acetone (usually 1:1 v/v), are the most common solvents reported in the literature for the extraction of adipates, phthalates and citrates from various food substrates. Various alkanes tested (e.g. pentane, hexane, heptane) performed adequately for the extraction. Also, several mixtures of hexane/acetone of different polarity were tested. It was obvious from the chromatograms obtained, given that multiple new peaks appeared in the area of interest, that acetone enhances dissolution of the fat matter. This conclusion, along with the high extraction efficiency of *n*-hexane, has been stressed by several researchers (Nerin et al., 1992; Wei et al., 2009). Thus, hexane was chosen as the extraction medium.

The most common methods with respect to solvent extraction of plasticizers from foodstuffs reported in the literature involve (a) use of a Soxhlet apparatus, (b) solvent extraction by homogenization, and (c) application of ultrasonics or microwaves (Badeka et al., 1999; Cano et al., 2002; Goulas and Kontominas, 1996; Goulas et al., 2007; Kondyli et al., 1992; Mercer et al., 1990; Nerin et al., 1992; Petersen and Breindahl, 2000; Startin et al., 1987). Soxhlet extraction needs to be combined with a clean up step; it is laborious, time-consuming and cannot be applied to multiple samples simultaneously. In addition, direct chromatographic analysis of the organic extract obtained by sample homogenization is unsatisfactory due to significant fat interference. It is reported in the literature that microwaves are mainly applied when solvents with polar groups are involved; non-polar solvents, as well as fat, which have a small relative dielectric constant, absorb to a lesser extent the energy of microwaves (Lau and Wong, 1996). A series of experiments showed that ultrasonic's yield maximum recoveries along with minimum fat interference; hence, ultrasonication was chosen as the extraction method.

Effect of Sample Size

In order to assess the effect of the sample size, an experiment was conducted by dividing the 45 g (6cm x 4cm x 1.3 cm) fish fillet samples into portions of a smaller size. No significant differences were observed in terms of recovery. Instead, a larger surface of substrate was exposed to the solvent allowing for greater fat dissolution to take place; hence, a clean-up step would be necessary in such case. Besides, in our previous study (Goulas et al., 2007) we reported that ATBC penetration into fatty foodstuffs does not exceed 7.5 mm (in depth) even in foods of high fat content (30% w/w) and relatively high temperatures (25°C). Given the present experimental conditions (skin with low fat content, low storage temperature), one can assume penetration to a lesser extent, thus no need exists to divide the 45 g sample into sub-samples.

Analytical Features of the Method

The optimized analytical parameters are summarized in Table 1. In order to evaluate the accuracy of the method, recovery tests were carried out as described under Section 2.6. Recoveries ranged from 80 to 95% for cod at all spiking levels tested, while for herring they were somewhat lower (with all individual values still exceeding 70%) especially at high plasticizer concentrations. This was attributed to the higher fat content of herring compared to that of cod (Table 2), along with the lipophilic nature of ATBC.

Table 1. Analytical parameters for ATBC plasticizer determined by GC/FID.

Parameter	ATBC
Retention time (min)	7.48 ± 0.05
LOD[a] (mg/kg)	1.6
LOQ[b] (mg/kg)	5.2
RSD[c] (%)	4.8
Regression equation	$E = 44.7 \times 10^{-3}$ C (mg/kg) – 12.5×10^{-2} [d]
Correlation coefficient (r)	0,9998

[a] Limit of detection, defined as three times the signal- to- noise ratio
[b] Limit of quantitation, defined as ten times the signal- to- noise ratio
[c] Relative standard deviation obtained from two extractions and three injections each (*n*=6)
[d] Peak area (arbitrary units)

Lipid and Moisture Analysis

Lipid and moisture content of muscle (g/100g muscle) of the selected fish species are summarized in Table 2. Cod was chosen as a typical non-fatty fish, while herring was chosen as a typical fatty fish.

Table 2. Lipid and moisture content (%) of muscle of cod and herring species.

Parameter	Mean value (%)	
	Cod	Herring
Lipid	0.7±0.1	13.2±0.4
Moisture	82.0±0.7	67.0±0.5

Values represent the mean of six determinations ± SD

Migration Results

The mean migration levels of ATBC into both fish species monitored as a function of time are shown in Tables 3 and 4. Equilibrium conditions were reached after ca. 6 days of contact for non-irradiated as well as electron processed samples of both marine fish species studied. Results indicate that the energy transferred to the polymer by ionizing radiation does not accelerate the attaining of equilibrium state (kinetics).

The final concentrations of ATBC in cod fillets ranged from 11.1 to 12.8 mg/kg (0.11–0.13 mg/dm^2), representing losses from the film between 1.0 and 1.1% of the available plasticizer. Respective values for herring samples were 32.4–33.4 mg/kg (0.32–0.33 mg/dm^2), corresponding to losses of ca. 2.9–3.0% of the available plasticizer. Data showed that irradiation with high energy electrons at pasteurizing doses did not considerably affect the film's specific migration characteristics. Statistically significant differences ($p < 0.05$) were only observed between non-irradiated and irradiated at 10 kGy cod fillet samples. On the contrary, statistically significant differences in plasticizer migration were found between the two fish species examined. Thus, fat content of the packaged fish fillets substantially affected the extent to which migration of ATBC occurred.

The amount of ATBC that migrated into fish samples cannot be discussed in relation to EU proposed upper limit for specific migration, since currently no SML exists for the specific additive (EC, 2009). Nevertheless, there are certain restrictions for additives for which a TDI or an ADI value has not yet been established. Contamination levels reached by the end of the storage period were far below the legal limit set for global migration (60 mg/kg or 10 mg/dm^2). On the other hand, ATBC concentrations in cod fillets exceeded the 5 mg/kg restriction as soon as after 24 hr of contact; for herring this value was exceeded after 12 hr of contact. Such high migration levels (as expressed in mg/kg units) are attributed to the high ratio of film surface to weight of food used in the present study in order to check compliance of the specific film with EU legislation under conditions which represent realistic use (i.e. food overwrapping or rewrapping). Grob et al. (2007) reported that for small packs with a high ratio of contact surface area to volume, present European legislation tolerates extremely high migration levels in terms of concentration in the food. The OM and SM of certain additives, for example DEHA from PVC cling films into cheese, demonstrate that such high migration values are realistically encountered (Castle et al., 1988b; Goulas et al., 2000; Petersen et al, 1995). Therefore, the authors propose that migration limits should be re-defined and expressed as migration amount per contact surface area taking into account a surface to weight ratio of 20 dm^2/kg rather than the presently used ratio of 6 dm^2/kg.

Table 3. Migration values of ATBC plasticizer from control and electron-irradiated saran film into cod fillets at 4 ± 1°C

Contact time (hr)	ATBC migration[a]								
	Non- irradiated			5 kGy			10 kGy		
	((mg/kg)	(mg/dm^2)	% loss	(mg/kg)	(mg/dm^2)	% loss	(mg/kg)	(mg/dm^2)	% loss
12	2.3±0.1	0.02±0.001	0.21±0.01	2.4±0.1	0.02±0.001	0.21±0.01	3.0±0.1	0.03±0.001	0.26±0.01
24	5.6±0.3	0.06±0.003	0.50±0.02	5.8±0.3	0.06±0.003	0.51±0.02	6.7±0.3	0.07±0.002	0.60±0.02
48	7.7±0.4	0.08±0.004	0.69±0.03	8.4±0.4	0.08±0.002	0.75±0.04	9.3±0.4	0.09±0.004	0.83±0.04
96	9.8±0.5	0.10±0.003	0.87±0.04	9.9±0.5	0.10±0.005	0.88±0.03	10.8±0.5	0.11±0.004	0.96±0.05
144	11.1±0.5	0.11±0.005	0.99±0.05	11.2±0.5	0.11±0.003	0.99±0.05	12.0±0.6	0.12±0.006	1.06±0.03
192	11.6±0.6	0.12±0.004	1.03±0.03	12.0±0.6	0.12±0.005	1.06±0.04	12.8±0.6	0.13±0.004	1.14±0.04
240	11.4±0.5	0.11±0.005	1.01±0.03	12.0±0.6	0.12±0.005	1.06±0.04	12.8±0.6	0.13±0.006	1.14±0.04

[a] Values represent the mean of two extractions and triplicate runs ± SD.

Table 4. Migration values of ATBC plasticizer from control and electron-irradiated saran film into herring fillets at 4 ± 1°C.

Contact time (hr)	ATBC migration [a]								
	Non-irradiated			5 kGy			10 kGy		
	((mg/kg)	(mg/dm^2)	% loss	(mg/kg)	(mg/dm^2)	% loss	(mg/kg)	(mg/dm^2)	% loss
12	6.5±0.3	0.06±0.003	0.57±0.03	6.8±0.3	0.07±0.003	0.60±0.03	7.3±0.4	0.07±0.003	0.65±0.03
24	18.6±0.9	0.19±0.009	1.65±0.08	18.6±0.9	0.19±0.009	1.65±0.08	19.5±0.9	0.20±0.010	1.73±0.08
48	24.4±1.2	0.24±0.012	2.17±0.10	24.1±1.2	0.24±0.012	2.14±0.10	24.9±1.2	0.25±0.012	2.21±0.11
96	28.4±1.4	0.28±0.013	2.52±0.11	28.9±1.4	0.29±0.013	2.57±0.12	29.7±1.4	0.30±0.013	2.64±0.11
144	32.4±1.6	0.32±0.013	2.88±0.12	32.9±1.6	0.33±0.015	2.93±0.14	33.4±1.6	0.33±0.016	2.97±0.13
192	33.3±1.6	0.33±0.015	2.96±0.13	33.0±1.6	0.33±0.015	2.93±0.13	33.4±1.6	0.33±0.015	2.97±0.13
240	32.7±1.6	0.33±0.015	2.91±0.11	32.9±1.6	0.33±0.016	2.92±0.13	33.4±1.6	0.33±0.016	2.97±0.14

[a] Values represent the mean of two extractions and triplicate runs ± SD.

Based on existing data, a realistic consideration for a significant part of the population would be: For a 60 kg adult, a daily consumption of 150–200 g fish packaged in this type of film would result in a maximum daily plasticizer intake equal to 0.04 and 0.11 mg ATBC/kg body weight for cod and herring fillets, respectively. Since no TDI value has yet been established for ATBC, one cannot evaluate these daily intake values in terms of toxicological safety. The fact, though, that an adult consumes, on a daily basis, a variety of fatty foodstuffs which may have been in contact with PVC or saran cling-films probably poses a safety concern. An estimated maximum daily intake of 1.5 mg/kg body weight has been reported for ATBC (MAFF, 1991); currently, higher daily intakes are expected due to an increased usage of VDC copolymers in microwave applications.

Data of the present study are comparable to those reported by other researchers. Till et al. (1982) reported a migration value of dioctyl adipate (DOA) plasticizer from PVC film into fish fillets equal to 2.4 mg/dm^2 (1.3% loss from film) after 7 days of contact at 4°C, while the respective migration value determined in lean beef was 1.4 mg/dm^2 (0.7% loss). This is a contradictory finding, given the fact that fat content of lean beef is significantly higher than that of fish fillet. Such differences may be attributed to the nature of fish lipids, which differ considerably from those of other foodstuffs, such as beef, poultry, pork, and so on. It is well known that fish fat contains a high percentage of polyunsaturated fatty acids and thus it is fluid.

Castle et al. (1988 b) reported migration values of ATBC (4.8% w/w) from saran film into cheese ranging from 1.3 to 7.7 mg/kg after 5 days of contact at room temperature. According to the same authors, migration levels of ATBC from the same film into cheese, cake and sandwich reached 6% (0.6 mg/dm^2), 2% (0.2 mg/dm^2), and 1% (0.1 mg/dm^2), respectively, after 5 days of contact at 5°C.

According to Goulas and Kontominas (1996), no statistically significant differences were found in DOA (28.3% w/w) migration from PVC film into chicken meat (7% fat) between irradiated (4 and 9 kGy of gamma radiation) and non-irradiated samples after a 10 day contact period at 4–5°C.

Goulas et al. (2000) studied the migration of DEHA plasticizer (28.3% w/w) from PVC film into hard and soft cheeses. The authors reported that after 240 hr of contact under refrigeration, the migration of DEHA was approximately 345.4 mg/kg (18.9 mg/dm^2) for Kefalotyri cheese, 222.5 mg/kg (12.2 mg/dm^2) for Edam cheese, and 133.9 mg/kg (7.3 mg/dm^2) for Feta cheese, corresponding to plasticizer losses from the film equal to 37.8%, 24.3%, and 14.6%, respectively. Migration values exceeded, in all cases, the upper limit set by the EU for DEHA (18 mg/kg or 3 mg/dm^2). In the above study, the ratio of film surface to weight of food was ca. 0.64 dm^2 to 35g cheese (18:1). According to the authors, the different fat and moisture content of the three cheese types tested resulted in statistically significant differences in plasticizer migration values.

It should be stressed at this point that migration levels (as expressed in mg/dm^2 and percentage loss from film) of ATBC plasticizer to both fish species reported in the present study, are consistent with migration levels attained for low fat foods and may be attributed to a variety of factors including (a) low initial content of plasticizer into the film, (b) high compatibility of ATBC with vinyl resins, (c) small thickness of the saran film, (d) low storage temperature of fish samples, (e) low fat and high moisture content of fish samples, and (f) low fat content of fish skin (fat of the specific marine species is mainly located within the muscular tissue and the liver).

CONCLUSION

An analytical methodology for the extraction of additives of medium polarity, such as plasticizers, from complex food matrices was proposed. An effort was made to eliminate the clean up step necessary for the separation of the target analyte from the co- extracted fat matter. The application of such an analytical approach allows for (a) efficient extraction of ATBC plasticizer at ppm levels, and (b) determination of the analyte using GC with no impairment of the separation efficiency of the column due to fat interference. Moreover, the method described is simple, rapid, and accurate and fulfils the requirement of applicability to multiple samples simultaneously.

By comparing ATBC migration values determined in both irradiated and control fish samples packaged in saran cling film, it is obvious that intermediate doses (≤ 10 kGy) of ionizing e-beam radiation do not significantly affect the copolymer in terms of its specific (plasticizer) migration behaviour. On the contrary, fat content of the packaged foodstuff substantially affects the total amount of migrating additive.

Two issues to be noted at this point are (a) the legal issue of non- compliance of the saran film used with EU restrictions and (b) the issue of public health. As also reported by other researchers (Castle et al., 1988b; Goulas et al., 2000; Grob et al., 2007; Petersen et al., 1995), high migration levels in terms of concentration in the food (mg/kg) are often encountered as a result of food overwrapping with flexible films. It is, thus, suggested that present migration limits should be re-examined and redefined on a different basis.

ACKNOWLEDGMENTS

This study was funded by the General Secretariat of Research and Technology (GSRT) of Greece within the framework of the *PENED 2001* Program of the Greek Ministry of Development. The authors gratefully acknowledge this financial support. Furthermore, the authors wish to acknowledge technical assistance provided by Dr. M. Stahl at the Federal Research Center for Nutrition and Food (Germany).

KEYWORDS

- **Acetyl tributyl citrate**
- **Fish fillets**
- **Migration**
- **Polyvinylidene chloride/polyvinyl chloride**

Chapter 5

Evaluation of Bag-In-Box as Packaging Material for Dry White Wine

M. Revi, A. Badeka, and M.G. Kontominas

INTRODUCTION

The objective of the present study was to evaluate bag-in-box as packaging material for white dry wine of the VILANA variety. Two types of composite bags were tested: (1) with an inner layer of low density polyethylene (LDPE) and (2) with an inner layer of ethylene vinyl acetate (EVA). Samples were stored at 20°C in the dark for 6 months. The sampling was carried out at 3, 30, 60, 90, and 180 days. Glass bottles 1.5 l in capacity was used as controls. The following quality parameters were monitored: Total acidity, volatile acidity, pH, total SO_2, free SO_2, color (absorbance at λ = 420 nm), ascorbic acid content, semi-quantitative determination of volatile compounds using Solid Phase Microextraction—Gas Chromatography (SPME-GC/MS) and sensory evaluation. Focus was placed on possible flavor scalping of volatiles by the packaging material.

The bag-in-box multilayer packaging material affected the total SO_2, free SO_2, ascorbic acid content and color, most probably due to its low oxygen permeability as compared to glass. Volatile analysis showed that a considerable part of wine aroma compounds were absorbed by the plastic, with organic acids exhibiting the highest sorption potential. Between the two plastics used as inner layers of the multilayer bag, LDPE showed a higher sorption tendency. In contrast, wine packaged in glass retained the largest part of its volatile compounds.

The sensory evaluation showed that shelf life of white wine packaged in multilayer bag-in-box type materials does not exceed 3 months while wine packaged in glass bottles retained acceptable quality for at least 6 months.

The quality of wine is a function of product composition and organoleptic properties such as color, body and most of all flavor. Wine is composed of water (80–85%), alcohols (9–15%) and minor constituents (~3%) (Jackson, 2000). Minor constituents include organic acids, sugars, phenols, nitrogen compounds, enzymes, vitamins, lipids, inorganic anions and cations, and a large number of flavor compounds. Of these, organic acids and flavor compounds are of prime importance substantially affecting product quality. Major organic acids include tartaric (2–5 g/l), malic (0–4 g/l), and citric acid (0–4 g/l). Tartaric acid and its salts are responsible for wine total or titrable acidity while acetic acid ($\leq$ 2 g/l) is mostly responsible for wine's volatile acidy (Amerine and Ough, 1974).

Wine aroma is usually a complex mixture of volatile compounds belonging to higher alcohols and organic esters. However aldehydes, ketones and terpenes also contribute to wine aroma (Jackson, 2000)

Wines develop their aroma from aromatic compounds in grapes (primary aroma), from the fermentation process (secondary aroma) and from flavor compounds formed as a result of ageing (tertiary aroma, bouquet). (Jackson, 2000).

Phenolic compounds are responsible for the characteristic color, astringency and antioxidant activity of wines and include phenolic acids, flavonoids, anthocyanins, and tannins (Amerine and Ough, 1974).

The packaging aims to protect and retain the original quality of foods and beverages (Nielsen and Jägerstad, 1994). Key parameters forward the fulfillment of this objective are the degree of barrier of the packaging material with regard to oxygen, light and moisture (Robertson, 2006) and the inertness of the packaging material with regard to migration and flavor scalping (Amerine and Ough, 1974; Kontominas, 2010).

Besides glass several other flexible packaging materials have been used in wine packaging including polyethyleneteephthalate (PET), tetrabrick type and bag-in-box type containers (Roberston, 2006).

The bag-in-box container is composed of a multilayer bag usually containing metalized polyester as the barrier layer and an inner layer of either LDPE or EVA placed inside a rectangular paperboard container for mechanical protection purposes. The bag is equipped with a special polypropylene tap for wine serving. Bags are filled under vacuum and the headspace is filled with an inert gas (N_2).

As wine is removed through the tap, the bag collapses protecting the remaining product from the effect of oxygen. This type of packaging is popular for table wines all around the world in capacities between 5 and 20 l. One problem associated with bag-in-box containers is that the polyolefins and their copolymers used as inner layers are known to be potent absorbers of volatile flavor compounds, being hydrophobic in nature (Salame, 1989). Thus beverages such as citrus juices and wine are expected to lose a substantial part of their aroma due to sorption. Halek and Luttman (1991) reported that between carvone ($C_{10}H_{14}O$) and d-limonene, two similar terpenes of different polarity, d-limonene was adsorbed LDPE much more rapidly than carvone.

Orange juice volatiles were adsorbed to different degree by LDPE following the sequence hydrocarbons > ketones > esters > aldehydes > alcohols (Linssen and Roozen, 1994; Linssen et al., 1991; Nielsen et al., 1992).

Based on the above, the objective of the present study was to evaluate bag-in-box as a packaging material for white wine of the VILANA variety. Analysis of physicochemical parameters and sensory evaluation were carried out over a period of 12 months at room temperature.

MATERIALS AND METHODS

Dry white table wine of the VILANA variety was donated by the Mihalaki winery SA in Iraklion, Crete, Greece in November, 2008. It was produced from the same year's vintage. Bag-in-box pouches, 5 l in capacity, having an inner layer of LDPE or EVA respectively were donated by VLACHOS SA (Athens, Greece). Pouches were commercially filled and placed inside paperboard cartons so as to provide access to the plastic tap. Wine packaged in dark green glass bottles, 1½ l in capacity was used as the control sample. Packaged wine samples were stored in a controlled temperature cabinet (20°C) in the dark. Sampling was carried out at 3, 30, 60, 90, and 180 days.

Determination of total acidity, volatile acidity, total SO_2, free SO_2 was carried out according to AOAC 1996. The pH was determined by direct submersion of a Crison model 507 (Crison Instr., Barcelona, Sapin) pH meter into the wine. Color was determined by measuring the absorbance of wine at $\lambda = 420$ nm. Ascorbic acid was determined according to the Official EU method (EEC, 1990): 50 ml of wine were transferred to a 250 ml Erlenmeyer flask. The 5 ml of acetaldehyde solution were added so as to bind the SO_2 present in the wine. The sample was left for 30 min and then titrated with a N/50 I_2 solution using starch as indicator. The Ascorbic acid (AA) was calculated using equation: AA (mg/l) = n × 35.2, where n = ml of N/50 I_2 solution consumed.

Semi-Quantitative Determination of Volatile Compounds

Solid Phase Microextraction (Spme) Sampling

A volume of 2 ml of wine, 0.5 g NaCl and 20 μl of internal standard (4-methyl–2-pentanol) were placed in a 4 ml glass vial and sealed with a screwed-cap equipped with a Teflon-coated needle-pierceable septum. The SPME fiber used was coated with 50/30 μm divinylbenzene/carboxen on polydimethylsiloxane. (Supelco, Bellefonte, PA, USA). The vial was placed in a 40°C water bath and equilibrated for 5 min. After equilibration the needle of the SPME holder was inserted into the vial through the septum and the fiber was exposed to the headspace of the sample for 20 min to adsorb the volatiles. Subsequently, the fiber was retracted into the needle assembly and transferred to the injection port of the GC/MS unit.

Gas Chromatography–Mass Spectrometry Analysis

A Hewlett-Packard 6890 series gas chromatograph equipped with a Hewlett-Packard 5973 mass selective detector (Wilmington, DE, USA) was used for the analysis of volatile compounds adsorbed onto the SPME fiber. The column used was a DB-5 MS (60 m × 0.320 mm i.d and 1 μm film thickness, J & W Scientific, Agilent Technologies, Folsom, USA). The flow rate of the helium carrier gas was 1.5 ml/min. The injector temperature was 260°C in split mode (2:1). The SPME fiber remained in the injector for 2 min. The initial temperature of column was 40°C held for 5 min, then raised to 260°C at a rate of 5°C/min and held at 260°C for 5 min. The MS conditions were as follows: Source temperature: 230°C, Quadrupole temperature: 150°C; transfer line temperature: 280°C; acquisition mode electron impact (EI 70 eV) and mass range m/z: 30–350. Identification of volatile compounds was achieved by comparing the mass spectra of the recorded chromatographic peaks with the Wiley 275 MS data base while semi-quantification was achieved by comparing the MS detector response of the internal standard to that of the recorded peaks (J. Wiley & Sons Ltd., West Sussex, England).

Sensory Evaluation

Sensory evaluation (acceptability test) was carried out by 51 member untrained panel (20 males and 31 females) consisting of faculty and graduate students of the Laboratory of Food Chemistry and Technology of the Department of Chemistry, University of Ioannina. Panelists were chosen using the following criteria: Ages between 22 and 60, non-smokers, who consume wine regularly. Judges were asked to evaluate appearance, odor and taste of wine on a scale between 5 and 1 where 5 = most liked sample equal to that of the primary control sample and 1 = least liked sample (unacceptable). The primary control sample consisted of wine packaged in dark green glass bottles and held in the dark at 13°C for a period of 6 months. Before evaluation the primary control sample was left to equilibrate at room temperature. It was then presented to judges along with the three test samples at each sampling period.

Statistical Analysis

Data were subjected to analysis of variance (ANOVA) using the software SPSS 16 for windows. Means and standard error were calculated, and, when F-values were significant at the $p < 0.05$ level, mean differences were separated by the least significant difference procedure.

DISCUSSION AND RESULTS

Determination of Titrable Acidity

Titrable acidity values expressed as tartaric acid as a function of packaging material and storage time at 20°C is given in Table 1.

Table 1. Titrable acidity values of VILANA white wine.

	Tartaric acid (g/l) Mean values ± SD		
Storage days (20°C)	Glass	Bag with inner layer LDPE	Bag with inner layer EVA
3	5.50±0.21	5.50±0.12	5.50±0.07
30	5.58±0.05	5.50±0.05	5.60±0.06
60	5.45±0.05	5.51±0.06	5.51±0.01
90	5.46±0.06	5.55±0.04	5.58±0.05
180	5.54±0.01	5.70±0.07	5.77±0.04

Titrable acidity of VILANA white wine was not affect by the packaging material up to day 90 of storage. On day 180 wine packaged in composite bags internally lined with polyethylene (PE) or EVA showed a small but statistically higher ($p < 0.05$) acidity value. Such a small increase in titrable acidity may be attributed either to the very small oxygen permeability of the composite bag material or to lack of complete air tightness of the tap. Titrable acidity values were close to but within the regulatory limit value of 8.0 g/l (EEC, 1990).

Determination of Volatile Acidity

Volatile acidity values expressed as acetic acid, as a function of packaging material and storage time are given in Table 2.

Table 2. Volatile acidity values of VILANA white wine.

	Acetic acid (g/l) Mean values ± SD		
Storage days (20°C)	Glass	Bag with inner layer LDPE	Bag with inner layer EVA
3	0.32±0.01	0.32±0.01	0.32±0.01
30	0.32±0.01	0.32±0.01	0.32±0.01
60	0.33±0.00	0.33±0.01	0.31±0.01
90	0.33±0.01	0.34±0.01	0.31±0.01
180	0.33±0.01	0.34±0.07	0.32±0.00

Volatile acidity of VILANA white wine was not affected by both the packaging material and storage time. Volatile acidity values were close to but within the regulatory limit values of 0.35 g/l (EEC, 1990).

Determination of pH

The pH depends on the nature and the concentration of organic acids in wine. Its importance is greater than that of total acidity since the latter is only dependent on amount of organic acids. In contrast pH refers to the amount of free carboxyl groups giving H^+ through dissociation. The pH values of VILANA white wine as a function of packaging material and storage time are given in Table 3. The pH values of VILANA wine were not affected by both packaging material and storage time.

Table 3. The pH values of VILANA white wine.

	Mean values ± SD		
Storage days (20°C)	Glass	Bag with inner layer LDPE	Bag with inner layer EVA
3	3.36±0.03	3.36±0.03	3.36±0.02
30	3.39±0.02	3.39±0.04	3.38±0.04
60	3.39±0.01	3.38±0.01	3.39±0.05
90	3.35±0.04	3.34±0.02	3.32±0.03
180	3.37±0.03	3.38±0.01	3.36±0.01

Total SO_2

The determination of total SO_2 is important for two reasons) to control its concentration (≤210 mg/l for white wines) having been accused of various health related problems (McCoy, 2010 a and b) to monitor its concentration during wine storage since

SO_2 has an antimicrobial and thus preservative effect on wine. Total SO_2 values for VILANA white wine as a function of packaging material and storage time are given in Table 4.

Table 4. Total SO_2 values of VILANA white wine.

	Total SO_2 (mg/l) Mean values ± SD		
Storage days (20°C)	**Glass**	**Bag with inner layer LDPE**	**Bag with inner layer EVA**
3	149.10±2.69	135.35±4.53	131.90±0.85
30	137.60±2.97	106.20±0.85	108.15±2.76
60	123.20±1.84	99.85±0.92	98.55±2.76
90	118.10±2.40	92.15±3.61	92.45±1.34
180	110.10±3.25	73.90±0.42	84.00±3.39

It is noteworthy to mention that as early as 3 days of storage the amount of total SO_2 was different ($p < 0.05$) in glass versus composite plastic bags. In all cases total SO_2 content decreased with storage time. Decrease was substantially higher ($p < 0.05$) in the case of the two composite plastic bags. Losses in total SO_2 were 26.1%, 45.4%, and 36.5% for glass, PE lined, and EVA lined composite bag. Higher losses of total SO_2 in the two composite bags may be attributed to the low oxygen permeability of the plastic packaging material and possibly lack of complete air tightness of the tap. Loss of total SO_2 in the glass bottled wine may be related to the oxygen either contained in the glass headspace or to the oxygen dissolved in the wine.

Determination of Free SO_2

Free SO_2 is in the form of SO_2, H_2SO_3 or sulfurous salts with cations in wine. The SO_2 is the major antimicrobial substance in wine possessing antioxidant substrates from oxidation. Free SO_2 values for VILANA white wine as a function of packaging material and storage time are given in Table 5.

Table 5. Free SO_2 values of VILANA white wine.

	Free SO_2 (mg/l) Mean values ± SD		
Storage days (20°C)	**Glass**	**Bag with inner layer LDPE**	**Bag with inner layer EVA**
3	35.35±1.48	26.40±2.97	28.55±1.48
30	32.60±0.28	12.80±1.84	13.15±2.33
60	28.55±1.48	10.20±0.10	11.50±0.42
90	26.10±0.85	6.30±0.06	7.35±0.50
180	24.10±0.71	3.80±0.09	5.8±0.14

As with total SO_2, a substantial loss ($p < 0.05$) in free SO_2 was noted in both composite bags as compared to glass as early as 3 days of storage. At the beginning of storage both composite bags showed the same free SO_2 content. However, after 60 days and during the rest of storage the EVA lined composite bag retained a significantly ($p < 0.05$) higher amount of free SO_2 as compared to the PE lined composite bag. Losses in free SO_2 were 32.0%, 86.0%, and 80.0% for glass, PE lined and EVA lined composite bags respectively. Present results are in agreement with those of Bach and Hess (1984) who reported similar losses in free SO_2 in white wine packaged in PET and PE bags.

Determination of Color

Absorbance at 420 nm is used as an index of spoilage of wines due to oxidation (Godden et al., 2001; Skouroumounis et al., 2003). Absorbance values at 420 nm as a function of packaging material and storage time as given in Table 6.

Table 6. Absorbance at 420 nm of VILANA white wine.

	Mean values ± SD		
Storage days (20°C)	Glass	Bag with inner layer LDPE	Bag with inner layer EVA
3	0.070±0.000	0.075±0.000	0.075±0.001
30	0.076±0.002	0.084±0.003	0.080±0.000
60	0.083±0.000	0.090±0.000	0.087±0.000
90	0.091±0.003	0.100±0.000	0.095±0.001
180	0.100±0.007	0.126±0.002	0.118±0.000

As shown in Table 6 absorbance values increased for all packaging materials with storage time. Already after 3 days of storage, absorbance values of both composite bags are higher ($p < 0.05$) by ca 6.7% then those of glass. As storage time progresses degree of wine oxidation seems to be affected by the packaging material. After 180 days of storage absorbance values increased by 43% in glass, 57% in the EVA lined and 68% in the PE lined composite bags. Thus the EVA lined composite bag showed a somewhat better behavior with regard to protecting the wine from oxidation. It is noteworthy to mention that even wine packaged in glass and kept in the dark suffers oxidation due to the presence of both headspace oxygen and dissolved oxygen in the wine. Increase in absorbance values ($\lambda = 420$ nm) in white wine may be related to: 1) oxidation of wine phenolic substances, 2) reaction mechanisms involving acetaldehyde and glyoxylic acid 3) caramelization and Maillard reactions (Li et al., 2008)

Determination of Ascorbic Acid

Ascorbic acid content of VILANA white wine as a function of packaging material and storage time are given in Table 7.

Table 7. Ascorbic acid content of VILANA white wine.

	Mg/l Mean values ± SD		
Storage days (20°C)	**Glass**	**Bag with inner layer LDPE**	**Bag with inner layer EVA**
3	60.71±0.38	50.14±1.91	54.48±1.88
30	59.65±0.21	28.75±1.63	34.40±1.13
60	57.70±0.14	24.60±0.00	28.25±0.21
90	34.95±0.35	21.80±0.99	24.85±0.35
180	27.70±0.57	14.10±0.00	18.05±0.35

Data in Table 7 shows a drastic drop in ascorbic acid content in all packaging materials with storage time. This drop was substantially greater in the two composite bags. Of the two composite bag materials the EVA lined bag showed a higher retention of ascorbic acid. On day 30 of storage the ascorbic acid content of wine in glass has ca. double to that in the composite bags. As previously stated, the oxygen permeability of the composite bags and possibly lack of complete air tightness of the tap are responsible for ascorbic acid degradation in wine. At the same time, degradation of ascorbic acid was also noted in glass, however, to a lower degree due to headspace and dissolved oxygen in the wine. Losses in ascorbic acid after 180 days of storage were 45.5%, 66.9%, and 71.9% for wine packaged in glass, EVA lined and PE lined composite bag respectively. Present results are in general agreement with those of previous studies (Ayhan et al., 2001; Muratore et al., 2005; Sobek, 2003).

With regard to the protection of ascorbic acid, results are consistent with those of total and the SO_2 and absorbance values at 420 nm implying a somewhat better protection of EVA lined composite bag as compared to the PE lined composite bag.

Semi-quantitative Determination of Wine Volatiles

Volatiles identified in VILANA white wine are shown in Table 8.

The 32 volatiles identified belong to the following classes of compounds; alcohols, organic acids, esters, aldehydes, and terpenes. Of the alcohols, 3-methyl-1-butanol, 2-methyl-1-butanol and 2-phenyl ethanol are the main contributors to the alcohol content of white wine. Changes in wine alcohols as a function of packaging material and storage time are given in Table 9.

Table 8. Volatile compounds of VILANA white wine.

Compound	Retention time (min)
1-propanol	6.76
Acetic acid	7.68
Ethyl acetate	8.64

Table 8. *(Continued)*

Compound	Retention time (min)
Isobutanol	9.22
Ethyl Propionate	12.69
3-methyl-1-butanol	13.84
2-methyl-1-butanol	14.00
4-methyl-2-pentanol (i.s.)	14.55
2,3-butanodiol	16.02
Ethyl butyrate	16.61
Ethyl lactate	17.12
Furfural	18.15
Ethyl isovalerate	18.87
1-hexanol	19.47
Isoamyl acetate	19.79
Hexanoic acid	23.20
3-(methylthio) propanol	23.95
Ethyl hexanoate	24.46
Hexyl acetate	24.93
Limonene	26.09
2-ethyl furanate	26.52
Linalool	28.21
Methyl octanoate	28.84
2-phenyl-ethanol	28.99
Octanoic acid	30.02
Diethyl succinate	30.50
Diethyl octanoate	31.18
2-phenyl ethyl acetate	33.34
Ethyl nonanoate	34.18
Decanoic acid	35.90
Ethyl decanoate	37.03
Isoamyl octanoate	38.49
Ethyl laurate	42.22

The concentration of 1-propanol decreased ($p < 0.05$) in all samples with storage time. The 1-propanol originates from threonine and possesses sweet taste (Silva et al., 2000; Zoecklein et al., 1995). Large concentrations of 1-propanol are indicative of probable microbiological spoilage (Silva and Malvata, 1998; Versini, 1993). Losses in 1-propanol were higher ($p < 0.05$) in both plastic bags as compared to that in glass. This finding is in general agreement with Mentana et al. (2009) who observed a reduction in 1-propanol concentration in wine packaged in PET. Isobutanol originates from valine (Zoecklein et al., 1995) through the process of the formation of higher alcohols (3-methyl-1-butanol, 3-methyl-2-butanol, 1-propanol, 2-phenyl-alcohol and isobutanol) from yeasts during fermentation. It contributes to wine with pleasant aromatic notes (Silva et al., 2000). Concentration of isobutanol decreased ($p < 0.05$) with storage time. Losses in isobutanol were higher in both plastic bags as compared to that in glass. Between the two plastics, loss of isobutanol was higher in the PE lines as compared to the EVA lined plastic bag. After 180 days of storage losses in isobutanol were ca. 9.9% for glass, 20.6% for the EVA lined bag, and 24.2% for the PE lined bag respectively, most probably owed to sorption by the polymer. Given that isobutanol is a major contributor to the aroma of wine (Ferreira et al., 1995; Rapp and Mandery, 1986) its reduction in concentration affects sensory characteristics of wine. Isoamylalcohols (3-methyl-1-butanol, 2-methyl-1-butanol) comprise the major fraction of higher alcohols in mine. They are formed through determination and decarboxylation of the aminoacid isoleucine respectively (Boulton et al., 1996; Kana et al., 1988). Their concentration is of paramount importance in wine quality since low isoamyl alcohol concentrations have been correlated with "weak" aroma profile wines (Falque et al., 2001; Silva et al., 2000). All concentrations higher than 400 mg/l of iso amyl alcohols contribute objectionable odors in wine. 3-methyl-1-butanol concentration decreased in all samples as a function of storage time. After 180 days of storage loss of 3-methyl-1-butanol was 17.8% in glass. The respective value for 2-methyl-1-butanol was 13.1%. The sorption behavior of the two plastics (PE and EVA) was similar implying negligible flavor scalping for these compounds. Butane-2,3-diol is produced through the action of yeasts and bacteria. According to Ribereau and Peynand (1966), butane-2,3-diol is formed through bacterial decomposition of citric acid. It has a minor contribution to wine flavor with sweet and sour tones. From an initial value of 2.4 mg/l, butane-2,3-diol increased ($p < 0.05$) to 3.00 mg/l at the end of storage in wine packaged in glass. An analogous increase was observed in wine packaged in the two plastic bags. No flavor scalping was observed for butane-2,3-diol. The 1-hexanol is associated with grape variety (Cantagrel et al., 1997; Falque et al., 2001). It contributes to wine aroma adding coconut like tones. At concentrations higher than 20 mg/l it is responsible for grassy flavors (Cantagrel et al., 1997; Silva and Malcata, 1998). The 1-hexanol decreased ($p < 0.05$) in concentration from an initial value of 3.40 mg/l to 2.69 mg/l at the end of storage in wine packaged in glass. An analogous decrease in 1-hexanol occurred in wine packaged in the plastic bags. No flavor scalping was observed for 1-hexanol.

Table 9. Concentration of alcohols (mg/l) of VILANA white wine as a function of storage time and packaging material.

Alcohol	Packaging material	Days of storage 3	30	60	90	180	
1-propanol	Glass	1.32±0.04	1.36±0.05	1.28±0.04	1.25±0.03	1.19±0.03	↓ _
	PE	1.39±0.03	1.30±0.04	1.26±0.03	1.26±0.03	0.96±0.02	
	EVA	1.49±0.05	1.38±0.04	1.30±0.04	1.10±0.02	1.01±0.04	
Isobutanol	Glass	5.37±0.18	5.22±0.20	4.97±0.17	4.67±0.16	4.84±0.16	↓ ↓
	PE	5.29±0.20	4.82±0.18	4.73±0.15	5.22±0.17	4.01±0.13	
	EVA	5.40±0.17	5.62±0.19	5.46±0.17	4.47±0.15	4.29±0.12	
3-methyl-1-butanol	Glass	65.95±2.11	58.69±2.31	55.29±1.89	50.22±1.22	54.22±1.18	– –
	PE	65.46±1.98	52.94±1.57	52.91±1.21	54.20±1.54	50.67±1.41	
	EVA	64.46±2.00	60.30±1.99	57.86±1.75	56.59±1.70	52.11±1.62	
2-methyl-1-butanol	Glass	18.64±0.71	17.85±0.54	16.67±0.52	15.32±0.45	16.20±0.51	– –
	PE	18.56±0.65	15.89±0.50	15.84±0.49	16.42±0.48	15.23±0.42	
	EVA	18.43±0.61	17.27±0.51	16.88±0.51	16.23±0.52	15.73±0.48	
2,3 butanodiol	Glass	2.46±0.12	2.67±0.13	2.79±0.12	3.09±0.12	3.00±0.13	– –
	PE	2.44±0.13	2.52±0.12	2.16±0.11	2.95±0.13	3.17±0.17	
	EVA	2.44±0.12	2.68±0.12	2.48±0.13	2.89±0.12	2.99±0.11	
1-hexanol	Glass	3.40±0.15	3.42±0.14	2.92±0.11	2.68±0.11	2.69±0.10	– –
	PE	3.22±0.14	2.96±0.12	2.94±0.12	2.94±0.12	2.72±0.12	
	EVA	3.37±0.15	2.41±0.13	2.39±0.10	2.55±0.09	2.49±0.08	
3-(methylthio)-propanol	Glass	0.21±0.01	0.22±0.01	0.23±0.01	0.15±0.01	0.09±0.01	– –
	PE	0.21±0.01	0.21±0.01	0.15±0.01	0.15±0.01	0.10±0.01	
	EVA	0.17±0.01	0.17±0.01	0.17±0.01	0.13±0.01	0.11±0.01	
2-phenyl-ethanol	Glass	22.40±0.80	27.61±0.95	25.76±0.84	28.47±0.98	29.68±0.91	↓ ↓
	PE	22.46±0.81	27.40±0.90	26.82±0.85	24.83±0.85	23.82±0.82	
	EVA	20.33±0.79	25.39±0.84	26.32±0.88	27.11±0.82	26.78±0.84	

At concentrations ca. 2 mg/l, 3-methyl-thio-propanol contributes to wine aroma with potato/cauliflower like tones (Meilgaard, 1981). Its concentration decreased ($p < 0.05$) with time in all packaging materials. Such a decrease may be attributed to its oxidation (Loyaux et al., 1981). The packaging material did not seem to have a direct effect on its concentration. The 2-phenylethanol is formed from phenylalanine during fermentation. It contributes to wine aroma with floral and rose tones (Falque et al., 2001; Monica-Lee et al., 2000; Silva et al., 2000). The 2-phenylethanol increased ($p < 0.05$) in concentration from an initial value of 22.40 mg/l to 29.68 at the end of storage in wine packaged in glass. Such an increase may be due to the hydrolysis of the respective ester (Bayouone et al., 2000). An analogous increase in 2-phenylethanol ($p < 0.05$) was observed in wine packaged in the two plastic bags. In contrast, Mentana et al. (2009) observed sorption of 2-phenyl ethanol in plastic containers made of PET.

Figure 1 highlights changes in alcohol concentration of wine after a 6 month storage period as a function of packaging material.

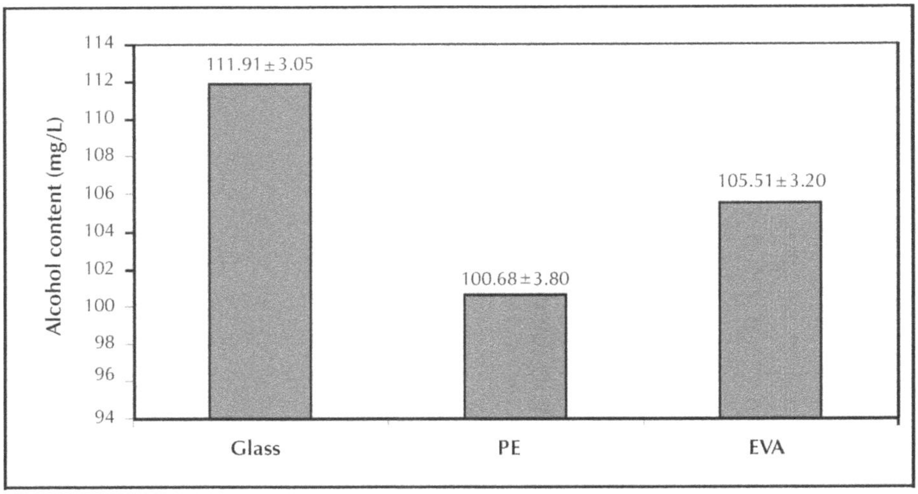

Figure 1. Alcohol content (mg/l) at 180 days of storage.

As shown in Figure 1, the largest concentration of alcohols was observed in wine packaged in glass. Both plastic bags showed significantly lower ($p < 0.05$) alcohol concentration. Differences between glass and the two plastic bags may be attributed either to flavor scalping or loss of volatiles through the packaging material. Given that both bags used were multilayer high barrier materials to organic vapors, oxygen and moisture, it is most probable that flavor scalping was responsible for the observed differences in concentration of alcohols. Of the two plastics, PE showed a higher scalping potential.

Changes of esters in wine as a function of packaging material and storage time are given in Table 10.

Significant changes in ester concentrations were observed. The majority of esters decreased in concentration ($p < 0.05$) with storage time with the exception of ethyl acetate, ethyl lactate, diethyl succinate and 2-ethyl furanate which increased ($p < 0.05$) in concentration and ethyl propionate which remained constant.

Table 10. Ester concentration (mg/l) of VILANA white wine as a function of storage time and packaging material.

Ester	Packaging material	Days of storage				
		3	30	60	90	180
Ethyl acetate	Glass	16.68±0.62	17.00±0.65	17.10±0.67	17.38±0.71	18.34±0.75
	PE	16.63±0.61	16.84±0.64	16.79±0.66	17.12±0.70	18.44±0.79
	EVA	16.14±0.62	16.37±0.65	16.47±0.65	17.76±0.93	18.03±0.77

Table 10. *(Continued)*

Ester	Packaging material	Days of storage 3	30	60	90	180
Ethyl propionate	Glass	0.22±0.007	0.24±0.007	0.23±0.006	0.20±0.006	0.22±0.007
	PE	0.22±0.007	0.24±0.008	0.22±0.007	0.20±0.006	0.22±0.006
	EVA	0.23±0.008	0.22±0.007	0.24±0.008	0.23±0.007	0.25±0.006
Ethyl butyrate	Glass	2.39±0.09	2.37±0.09	2.03±0.10	1.81±0.03	1.79±0.04
	PE	2.41±0.10	2.04±0.09	1.15±0.04	0.81±0.04	0.66±0.03
	EVA	2.42±0.11	2.78±0.11	1.30±0.03	1.13±0.04	1.05±0.03
Ethyl lactate	Glass	2.72±0.10	2.78±0.10	3.00±0.13	3.39±0.13	3.46±0.14
	PE	2.96±0.11	3.50±0.12	3.58±0.12	3.12±0.11	3.18±0.11
	EVA	2.80±0.10	3.69±0.12	3.43±0.13	3.57±0.12	3.33±0.12
Ethyl isovalerate	Glass	0.21±0.006	0.22±0.006	0.21±0.005	0.19±0.003	n.d.
	PE	0.21±0.005	0.18±0.004	0.21±0.004	0.20±0.003	n.d.
	EVA	0.23±0.005	0.19±0.005	n.d.*	n.d.	n.d.
Isoamyl acetate	Glass	9.53±0.21	8.82±0.22	6.91±0.21	4.90±0.13	4.05±0.15
	PE	9.72±0.25	7.34±0.21	7.04±0.21	5.34±0.16	3.01±0.14
	EVA	10.00±0.28	9.24±0.24	8.29±0.23	7.28±0.22	3.97±0.15
Ethyl hexanoate	Glass	30.04±1.15	26.04±1.10	25.86±0.94	25.59±0.95	25.29±0.92
	PE	32.53±1.18	23.70±0.96	22.87±0.83	22.82±0.81	22.26±0.80
	EVA	32.55±1.12	24.70±1.00	22.54±0.81	22.60±0.80	22.17±0.81
Hexyl acetate	Glass	0.64±0.024	0.57±0.022	0.42±0.021	0.29±0.007	0.23±0.007
	PE	0.64±0.023	0.43±0.020	0.39±0.020	0.23±0.007	0.11±0.005
	EVA	0.54±0.021	0.49±0.022	0.48±0.021	0.45±0.020	0.19±0.006
2-ethyl furanate	Glass	0.23±0.007	0.32±0.009	0.26±0.008	0.26±0.007	0.42±0.011
	PE	0.27±0.008	0.28±0.008	0.28±0.007	0.26±0.007	0.45±0.010
	EVA	0.26±0.007	0.25±0.007	0.29±0.009	0.32±0.009	0.41±0.010
Methyl octanoate	Glass	0.33±0.009	0.32±0.009	0.27±0.007	0.21±0.007	0.23±0.007
	PE	0.36±0.010	0.24±0.007	0.21±0.006	0.21±0.007	0.14±0.005
	EVA	0.41±0.020	0.33±0.010	0.37±0.010	0.37±0.011	0.23±0.008
Diethyl succinate	Glass	9.60±0.28	12.15±0.35	12.57±0.36	12.94±0.38	17.83±0.57
	PE	9.46±0.25	12.24±0.36	13.13±0.40	12.56±0.37	16.76±0.53
	EVA	9.48±0.29	11.82±0.35	14.46±0.42	14.28±0.49	19.60±0.64
Diethyl octanoate	Glass	86.29±3.00	75.23±2.88	74.97±2.89	67.64±2.48	68.71±2.40
	PE	84.62±3.10	73.05±2.78	68.45±2.50	65.79±2.45	53.79±2.12
	EVA	91.94±3.25	76.37±2.95	72.79±2.51	73.56±2.59	67.72±2.45
2-phenyl ethyl acetate	Glass	2.24±0.08	2.35±0.09	1.97±0.07	1.97±0.07	1.34±0.03
	PE	2.31±0.09	2.21±0.09	2.16±0.08	1.59±0.06	1.41±0.05
	EVA	2.35±0.10	2.21±0.08	2.15±0.08	2.09±0.07	

Table 10. *(Continued)*

Ester	Packaging material	Days of storage 3	30	60	90	180
Ethyl decanoate	Glass	0.22±0.007	0.21±0.007	0.14±0.005	0.12±0.004	0.14±0.005
	PE	0.17±0.005	0.14±0.005	0.13±0.005	0.13±0.003	n.d.
	EVA	0.23±0.006	0.20±0.005	0.17±0.005	0.17±0.005	0.10±0.003
Ethyl decanoate	Glass	32.54±1.10	27.36±1.00	20.83±0.81	20.65±0.77	20.39±0.75
	PE	17.39±0.68	8.97±0.35	9.49±0.36	7.37±0.33	4.19±0.17
	EVA	20.50±0.75	14.96±0.57	13.98±0.47	14.29±0.51	7.44±0.34
Iosamyl octanoate	Glass	0.32±0.009	0.26±0.007	0.15±0.002	0.15±0.004	0.15±0.003
	PE	0.34±0.010	0.22±0.007	0.24±0.006	0.17±0.005	0.06±0.003
	EVA	0.35±0.011	0.24±0.007	0.25±0.008	0.19±0.006	0.14±0.005
Ethyl lactate	Glass	0.71±0.027	0.49±0.016	0.41±0.015	0.39±0.012	0.36±0.010
	PE	0.10±0.003	n.d.	0.08±0.002	0.07±0.002	n.d.
	EVA	0.25±0.007	0.16±0.004	0.18±0.004	0.16±0.004	0.13±0.003

* non detectable

Ethyl acetate increased from 16.68 mg/l to 18.34 mg/l ($p < 0.05$) at the end of storage in glass containers. The same trend was observed for wines packaged in the two plastic bags. No effect of the packaging material was observed regarding ethyl acetate.

Ethyl lactate is formed during the malolactic fermentation from malic acid and ethanol (Maicas et al., 1999). It contributes to the aroma of wines and distillates with coconut and creamy tones. Ethyl lactate increased ($p < 0.05$) from 2.72 mg/l to 3.46 mg/l at the end of storage in wine packaged in glass. Such a trend was also observed for ethyl lactate by Cantagrel et al. (1997); Ferreira et al. (1997); Cantagrel et al. (1998) for wines and brandy during storage. The same trend was observed for wines packaged in the two plastic bags.

During the storage of wine, carbohydrate decomposition takes place. Decomposition products include 2-ethyl-furanate and furfural. Increase in 2-ethyl furanate ($p < 0.05$) during storage is most probably due to the above reaction mechanism. The specific contribution of 2-ethyl furanate to wine aroma has not been fully resolved. An analogous increase in its concentration was observed in all packaging materials tested.

Diethyl succinate showed a significant increase in wine during storage. In glass, diethyl succinate, increased ($p < 0.05$) from 9.60 mg/l to 17.83 mg/l at the end of storage. This trend is in agreement with that reported by Perez-Prieto et al. (2003) and Shinohara (1981). An analogous increase in diethyl succinate was observed in the two plastic bags.

Isoamyl acetate is a major contributor to the aromatic profile of wine (Van der Merwe and Van Wyk, 1981), decreased ($p < 0.05$) from 9.53 mg/l to 4.05 mg/l in glass packaged wine at the end of storage. This is in agreement with Perez-Coello et al. (2003) and Perez-Pireto et al. (2003). Both plastic bags behaved similarly to glass with regard to isoamyl acetate.

Ethyl propionate contributes with floral tones to wine aroma. Its concentration remained constant throughout storage. No considerable sorption of ethyl propionate was observed by plastic bags. In contrast, ethyl butyrate, bearing a longer carbon chain, showed a significantly higher concentration ($p < 0.05$) in wine packaged in glass as compared to that packaged in plastic bags. Of the two plastics, PE showed a higher scalping potential. Shimoda et al. (1988) reported that increase in the carbon chain length of esters resulted in a higher sorption potential.

Ethyl hexanate and ethyl octanate are major contributors to wine aroma (Ferreira et al., 1995; Rapp and Mandery, 1986). The concentration of both esters was substantially lower ($p < 0.05$) in the plastic bags as compared to glass. Sorption of ethyl hexanate is similar in both plastics, while that of ethyl octanate was higher ($p < 0.05$) in the PE lined bag. EVA's behavior resembles that of glass. Similar results were observed for hexyl acetate, methyl octanate, 2-phenyl acetate, ethyl nonanate and isoamyl octanate which showed a lower concentration ($p < 0.05$) in the PE lined bag as compared to the EVA lined bag and glass. It is postulated that an increase in the length of the ester carbon chain, results in a decrease in molecule polarity, leading to a higher sorption by the non-polar PE. Slightly polar EVA, in turn, exhibits a lower sorption potential. Along the same line of reasoning, decanoic ethyl ester and ethyl laurate exhibited the highest scalping potential even during the initial stages of storage (3 days) in plastic bags. Between the two, a higher ($p < 0.05$) sorption was observed in the PE lined plastic bag for decanoic ethyl ester. This finding is in agreement with that of Mentana et al. (2009) who reported a high sorption of decanoic ethyl ester by PET during the storage of wine. With regard to ethyl laurate, its concentration was reduced by 50% in glass and the EVA lined bag at the end of storage while in the PE lined bag it was non-detectable.

Figure 2 highlights changes in ester concentration of wine after a 6 month storage period as a function of packaging material.

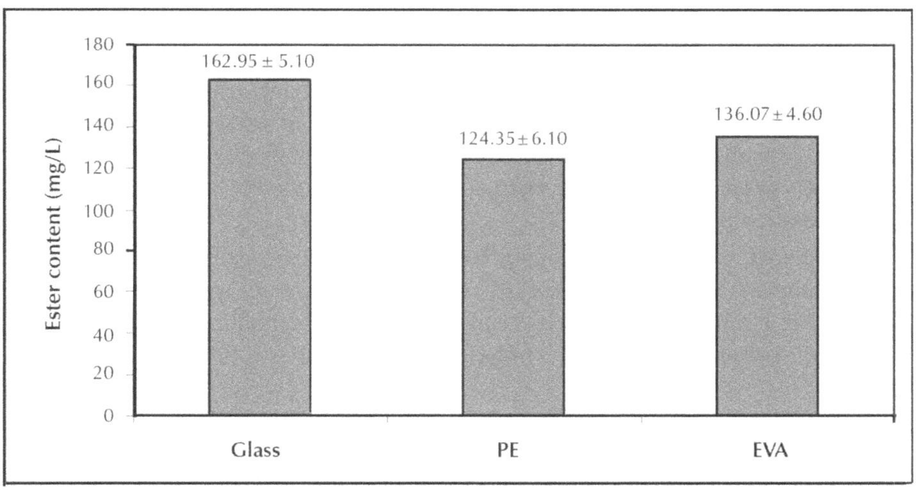

Figure 2. Ester content (mg/l) of VILANA white wine at 180 days of storage.

As shown in Figure 2 the largest concentration of esters was observed in wine packaged in glass. Both plastic bags showed significantly lower ($p < 0.05$) ester concentrations. Above trend may be rationalized as commented for alcohols.

Changes in organic acids concentration as a function of packaging material and storage time are given in Table 11.

Table 11. Acid concentration (mg/l) of VILANA white wine as a function of storage time and packaging material.

Acid	Packaging material	Days of storage 3	30	60	90	180
Acetic acid	Glass	2.26±0.08	2.17±0.08	2.40±0.10	2.42±0.10	2.49±0.11
	PE	2.69±0.12	2.68±0.11	2.48±0.10	2.40±0.10	2.74±0.12
	EVA	2.70±0.13	2.79±0.12	2.64±0.11	2.53±0.10	2.60±0.11
Hexanoic acid	Glass	3.70±0.14	4.42±0.15	4.48±0.14	4.23±0.13	4.71±0.16
	PE	3.64±0.12	4.13±0.16	4.14±0.14	3.26±0.13	3.23±0.12
	EVA	3.61±0.13	3.85±0.14	4.22±0.15	3.70±0.12	3.91±0.14
Octanoic acid	Glass	19.15±0.78	23.03±0.94	24.66±0.95	24.20±0.95	24.75±0.94
	PE	15.59±0.63	17.18±0.68	17.72±0.66	12.79±0.52	13.06±0.51
	EVA	16.63±0.65	17.17±0.65	17.53±0.66	15.18±0.61	17.53±0.67
Decanoic acid	Glass	4.14±0.15	4.21±0.14	4.17±0.16	4.41±0.16	4.77±0.17
	PE	0.88±0.03	0.40±0.014	0.31±0.012	0.41±0.014	0.15±0.012
	EVA	1.16±0.04	1.27±0.05	0.42±0.015	0.32±0.013	0.78±0.025

Wine packaged in glass showed a small but significant increase ($p < 0.05$) in organic acid concentration. This is in agreement with Hernandez et al. (2009) who reported an increase in hexanoic, octanoic, and decanoic acids during wine storage for a period of 12 months. It is interesting to note that decanoic acid concentration was drastically reduced as soon as day 3 of storage. At the end of storage concentrations for hexanoic, octanoic. and decanoic acids recorded were significantly lower ($p < 0.05$) in plastic bags as compared to glass. Between the two bags the PE lined bag showed the highest sorption potential.

Acetic acid is the product of either fermentation or wine spoilage during storage. Its concentration in glass increased ($p < 0.05$) from 2.26 mg/l to 2.49 mg/l while in both plastics it remained more or less constant. Figure 3 highlights changes in organic acid concentration of wine after 6 month storage period as a function of packaging material.

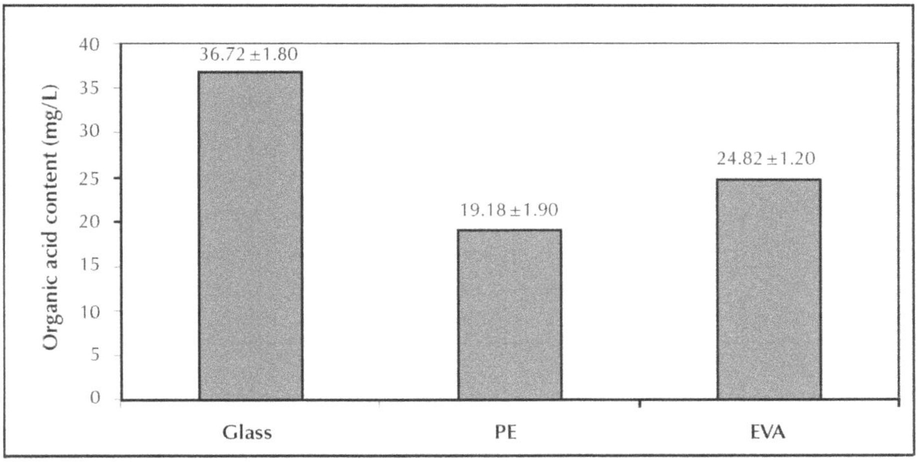

Figure 3. Organic acid content (mg/l) of VILANA white wine at 180 days of storage.

As shown in Figure 3 the largest concentration of organic acids was observed in wine packaged in glass. Both plastic bags showed significantly ($p < 0.05$) lower organic acid concentrations.

Changes in furfural concentration as a function of packaging material and storage time are given in Table 12.

Table 12. Furfural concentration (mg/l) of VILANA white wine as a function of storage time and packaging material.

	Packaging material	Days of storage 3	30	60	90	180
Furfural	Glass	0.27±0.01	0.23±0.01	0.27±0.01	0.36±0.02	0.95±0.03
	PE	0.23±0.01	0.28±0.01	0.26±0.01	0.23±0.01	1.02±0.03
	EVA	0.25±0.01	0.28±0.01	n.d.	0.26±0.01	0.97±0.02

Furfural is reported to contribute to the aroma of wines and distillates with dried nut tones (Monica-Lee et al., 2000). At high concentrations, furfural produces toxic effects in humans (Adams et al., 1997). According to numerous studies (Cole and Noble, 1997; Diez Marques et al., 1994; Perez-Coello et al., 1999; Mangas et al., 1996a, 1996b;) furfural content increases during aging of wines and distillates. Furfural concentration increased ($p < 0.05$) from 0.27 mg/l to 0.95 mg/l in wines packaged in glass at the end of storage. An analogous increase was recorded for wines packaged in plastic bags. Changes in terpenes' concentration as a function of packaging material and storage time are given in Table 13.

Table 13. Terpene concentration (mg/l) of VILANA white wine as a function of storage time and packaging material.

Terpene	Packaging	Days of storage				
	material	3	30	60	90	180
Limonene	Glass	0.17±0.007	0.15±0.005	0.15±0.005	0.14±0.004	0.13±0.005
	PE	0.18±0.006	0.16±0.005	0.15±0.004	0.16±0.005	0.09±0.001
	EVA	0.18±0.006	0.16±0.005	0.16±0.004	0.16±0.004	0.11±0.002
linalool	Glass	0.30±0.011	0.24±0.007	0.22±0.006	0.22±0.006	0.22±0.006
	PE	0.24±0.008	0.16±0.004	0.15±0.005	0.16±0.005	0.18±0.006
	EVA	0.26±0.006	0.26±0.005	0.27±0.006	0.26±0.008	0.20±0.005

The most extensively studied compound with regard to sorption is limonene. Limonene is an unsaturated terpene occurring in citrus fruits. It is a non-polar compound exhibiting high affinity to non-polar polymers (Moshanos and Shaw, 1998). At the end of storage, limonene concentration decreased ($p < 0.05$) from 0.17 to 0.13 mg/l. Lower limonene concentration values were recorded in the two plastic bags.

The PE showed the highest scalping potential for limonene. Present results are in agreement with those of Mahoney et al. (1988), Ikegami et al. (1991), Pieper et al. (1992) and Tawfik et al. (1998). Linalool is a terpene, originating from grapes. Due to its higher polarity as compared to limonene, it is less adsorbed by non-polar polymers such as PE and PP (Willige et al., 2002). Shimoda et al. (1988), Arora et al. (1991) and Charara et al. (1992) have shown that sorption by non-polar polymers increases with hydrophobicity of the sorbent. Decrease in linalool concentration during storage was similar for all three packaging materials tested. This finding was more or less expected based on the polar nature of linalool.

Figure 4 highlights changes in terpenes' concentration of wine after 6 month storage period as a function of packaging material.

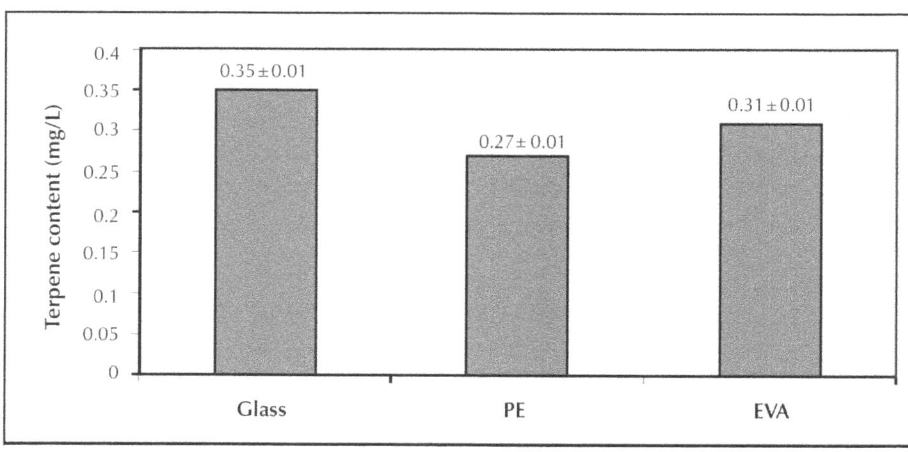

Figure 4. Terpene content (mg/l) of VILANA white wine at 180 days of storage.

As shown in Figure 4 the largest concentration of terpenes was observed in wine packaged in glass. Both plastic bags showed a significantly lower ($p < 0.05$) terpene concentrations due to sorption.

Figure 5 summarizes changes in total volatiles of wine after 6 months of storage as a function of packaging material. Wine samples packaged in glass showed the highest concentration of volatile compounds followed by the EVA lined plastic bag. Wine packaged in the PE lined bag retained the least amount of volatiles. For this packaging material sorption (scalping) followed the order; alcohols < terpenes < esters < organic acids. For the EVA lined bag the respective order was: alcohols < esters < terpenes < organic acids.

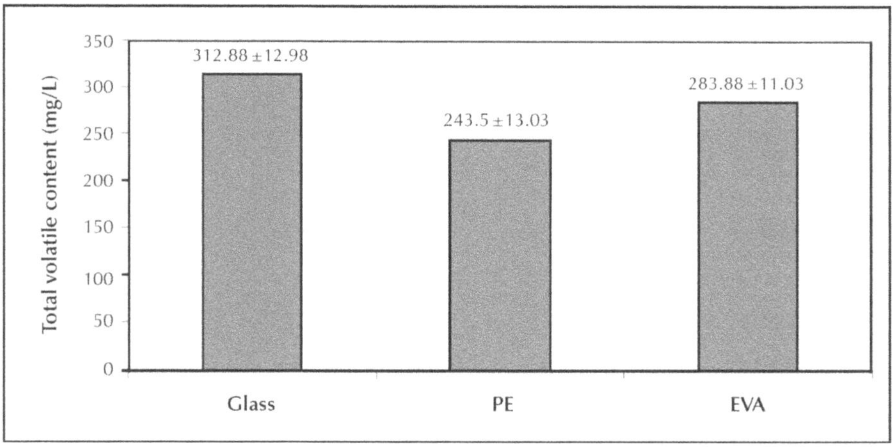

Figure 5. Total volatile content (mg/l) at 180 days of storage.

Sensory Evaluation

Sensory attributes were evaluated on day 3, 30, 60, 90, and 180 after bottling. Table 14 shows the results of the mean score of wine samples (including appearance, odor and taste) as a function of storage time and packaging material.

Table 14. Sensory evaluation of VILANA white wine.

Days of storage	Mean value ± S.D.		
	Glass	PE	EVA
3	5.00±0.00	4.79±0.22	4.90±0.15
30	4.46±0.32	3.59±0.68	4.06±0.43
60	4.13±0.78	3.47±0.51	3.43±0.57
90	3.89±0.19	3.22±0.38	3.27±0.10
180	3.94±0.27	2.37±0.62	2.51±0.40

Scoring scale 5 – 0 where 5 = excellent and 0 = unacceptable.

As shown in Table 14 as soon as 30 days after bottling, significant differences ($p < 0.05$) were observed between wine samples packaged in glass versus those packaged in PE lined bags. After 6 months of storage wine samples packaged in both plastic bags had substantially deteriorated as compared to samples packaged in wine. Based on a lower acceptability score of 3, it is obvious that wine in bag-in-box type of packaging becomes unacceptable after 90 days stored at room temperature. The same wine retains good sensory characteristics even after 6 months at the same temperature when packaged in glass and stored in the dark.

CONCLUSION

Deterioration in sensory quality of white wine (VILANA variety) is most probably enhanced by sorption of numerous volatile compounds such as isobutyl alcohol, hexanoic ethyl ester and octanoic ethyl ester by the polymeric packaging material. Between the two composite polymeric materials tested, the PE lined bag showed the highest potential for sorption. Glass is the most inert packaging material for wine. Experimentation is under way to extract and quantify sorbed volatiles from both PE and EVA polymeric films.

KEYWORDS

- **Bag-in-box packaging**
- **Quality retention**
- **Volatiles**
- **Wine**

ACKNOWLEGMENTS

The authors thank the Unit of Food Standardization, University of Ioannina, for providing access to the GC/MS instrument.

Chapter 6

Effect of Ozonation in Combination with Packaging and Refrigeration on Shelf Life Extension of Fresh Tomatoes

E.S. Karakosta and K.A. Riganakos

INTRODUCTION

Greenhouse tomatoes were collected from Preveza, Greece, washed, exposed to ozone (O_3) gas for 1 hr at various concentrations (0.5, 1 and 2 ppm) and stored at 4 ± 1°C for 49 days. Half of the samples were placed in commercial open crates and the other half were packaged in low density polyethylene (LDPE) bags. Peel firmness, color parameter values (L*, a* and b*), weight loss, total acidity, TVC, psychrotrophs, yeasts and moulds and volatile compounds were determined over the above 49 day storage period. In addition, sensory evaluation was carried out.

Results showed that changes in resistance of tomato fruit to penetration were statistically significant ($p < 0.05$) after day 7 for control, ozonated at 0.5 ppm, 0.5 ppm-LDPE and 1 ppm samples; after day 14 for 1 ppm-LDPE samples and after day 21 for 2 ppm and 2 ppm-LDPE samples. Ozonation dose significantly affected ($p < 0.05$) resistance to penetration after day 7 for all samples. The L* values were significantly affected ($p < 0.05$) by storage time in untreated samples. In contrast, ozonation dose did not significantly affect ($p > 0.05$) L* values. There were no statistically significant changes in a* and b* values ($p > 0.05$) either with storage time or ozonation dose.

Weight loss showed significant changes ($p < 0.05$) with storage time in all samples. Weight loss was significantly affected ($p < 0.05$) after day 7 for ozonated samples in crates and after day 0 for ozonated samples packaged in LDPE bags. Both storage time and O_3 concentration had a statistically significant effect ($p < 0.05$) on citric acid concentration for all samples after day 7.

Populations of mesophilic and psychrotrophic bacteria, yeasts and moulds were affected ($p < 0.05$) by ozonation after day 7. Storage time significantly affected ($p < 0.05$) TVC between 0 and 28 day, psychrotrophs between 0 and 35 day and yeasts/moulds between 0 and 49 day for ozonated samples ($p < 0.05$). Storage time significantly affected TVC, psychrotrophs and yeasts/moulds between 0 and 35 day for control samples.

There were no statistically significant changes ($p > 0.05$) recorded in most volatile-aromatic components of fresh tomato samples. Exceptions to this trend were ethanol which decreased with ozonation and increased with storage time, 1-penten-3-one, trans 2-pentenal, N-hexanal, trans 2-heptenal and geranyl acetone which increased and c is 3-hexenol which decreased in concentration with storage time.

Finally, sensory evaluation showed that the ozonated samples at concentrations of 0.5, 1 and 2 ppm packed either in commercial crates or in LDPE bags had better sensory characteristics compared to control samples. Texture, aroma, taste, and overall appearance scores of the fruit were kept above the lower acceptability limit in all ozonated samples.

Ozone (Figure 1) was first discovered by Christian Friendrich Schönbein in 1839. It is a strong oxidizing agent (Kogelschatz, 1988). At –112°C, O_3 gas condenses to a dark blue liquid (Oehlschlaeger, 1978). It is a triatomic molecule (O_3) that is considered to be an allotropic form of oxygen. It has a pungent, characteristic odor described as similar to "fresh air after a thunderstorm" (Coke, 1993). The O_3 has a relative molecular mass of 48, boiling point of –111.9°C and melting point of 192.7°C at 1 atm. The oxidation potential of O_3 is high (2.07 V) compared to that of hypochlorous acid (1.49 V) or chlorine (1.36 V) (Kim et al., 1999).

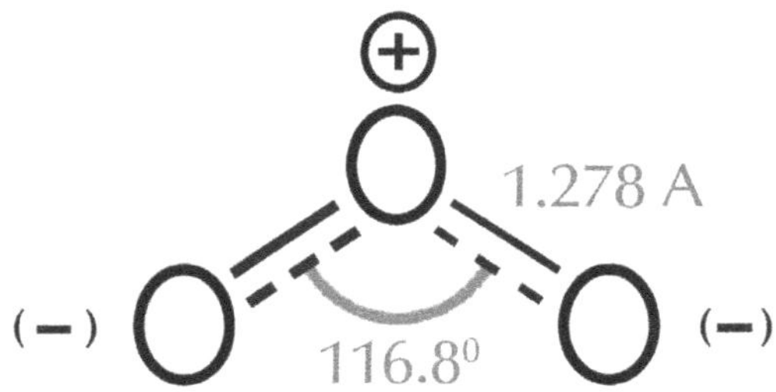

Figure 1. Ozone molecule.

The O_3 use has many advantages in the food industry. Suggested applications include: Food surface hygiene, sanitation of food plant equipment, reuse of waste water, lowering biological oxygen demand (BOD) and chemical oxygen demand (COD) of food plant waste (Dosti, 1998; Guzel-Seydim, 1996; Majchrowicz, 1998; Rice et al., 1982).

The bactericidal effects of O_3 have been studied and documented on a wide variety of organisms, including Gram positive and Gram negative bacteria as well as spores (Fetner and Ingols, 1956; Foegeding, 1985; Ishizaki et al., 1986; Restaino et al., 1995).

The O_3 destroys microorganisms by the progressive oxidation of vital cellular components. The bacterial cell surface has been suggested as the primary target of ozonation. Two major mechanisms have been proposed for the destruction of the target organisms. The first mechanism is oxidation of sulfhydryl groups and amino acids of enzymes, peptides and proteins to shorter peptides and the second mechanism is oxidation of polyunsaturated fatty acids to hydro peroxides (Victorin, 1992).

The O_3 has been approved in the US and classified as a food additive (US FDA, 2001). The decomposition of O_3 is so rapid in the aqueous phase of foodstuffs that its antimicrobial action is exhibited mainly at the food surface (Achen, 2000), leaving no residues (Guzel-Seydim et al., 2004a, 2004b). Recently, industry has indicated that the postharvest use of O_3 is increasing (Parish et al., 2003). In fact, in addition to its antimicrobial activity, O_3 is used in cold rooms to reduce the ethylene (C_2H_4) level in air and to extend the storage life of fruits and vegetables (Skog and Chu, 2000).

The O_3 has been proven effective in extending the shelf life of many fruits and vegetables, such as bananas (Gane, 1936), potatoes, onions and sugar beets (Baranovskaya et al., 1979), tomatoes and mandarins (Jin et al., 1989) and blackberries (Barth et al., 1995). Ewell (1950) showed that 2–3 μl/l of O_3 applied continuously or for a few hours per day doubled the storage life of strawberries, raspberries and grapes. More recently, Singh et al. (2002) showed that O_3 (combinations of 2.1, 5.2, or 7.6 mg / l for 5, 10, or 15 min) reduced the microbial counts in lettuce and carrots inoculated with *Escherichia coli.*

Continuous exposure to O_3 at 0.3 ppm increased moisture loss after 5 weeks of storage at 5°C and 90% relative humidity in peaches, but not after 4 weeks of storage in grapes (Palou et al, 2002). Nadas et al. (2003) stored strawberries for 3 days at 2°C in air with or without 1.5 ppm O_3 and then transferred fruits to room temperature. The O_3-treated fruits showed less weight loss than the non-treated samples after cold storage. However, differences were not significant at the later stages of storage at ambient temperature. Probably, O_3 treatment reduced moisture loss through transpiration of the fruit, but this effect diminished when the fruit was returned to ambient temperature. The weight loss subsequently increased.

Ozonated water treatment resulted in no significant differences in total sugar content of celery (Zhang et al., 2005). Perez et al. (1999) stored strawberries for 3 days at 2°C in an atmosphere containing 0.35 ppm ozone, and then transferred the fruit to 20°C for 4 days storage. Sucrose content of treated and non-treated fruits decreased with storage time. A fluctuation in glucose and fructose levels was observed from 0 to 5 day. The pattern of conversion of sucrose into glucose and fructose was significantly different in treated and non-treated samples. Low contents of sucrose, glucose and fructose were measured on the third day of storage. This could be due to an activation of sucrose degradation pathways in response to oxidative stress caused by O_3 (Karaca and Velioglu, 2007).

Skog and Chu (2001) reported that O_3 (0.4 μmol/mol) significantly improved the quality and storage life of cold-stored broccoli and cucumber. Barth et al. (1995) and Salvador et al. (2006) reported similar findings for O_3-treated blackberries and persimmon, respectively. Maguire and Solverg (1980) reported that tomatoes ripened in 3.7 μl/l O_3 had a more pronounced typical tomato aroma than those ripened in air. Tzortzakis et al. (2007) reported that O_3-treatment did not affect fruit weight loss, antioxidant status, CO_2/H_2O exchange, ethylene production or organic acid, vitamin C (pulp and seed) and total phenolic content when tomatoes (*Lycopersicon esculentum*) were exposed to O_3 concentrations ranging between 0.005 (controls) and 1.0 μmol / mol.

Naitoh and Shiga (1989) reported that simultaneous treatment with an ozone-air mixture (0.02–0.2 ppm) and ozonated water (0.3–0.5 ppm) decreased the total microbial count and elongation of hypocotyls of bean sprouts (black matre and alfalfa). Catalase and sureroxide dismutase activities increased significantly with O_3 treatment during germination. The effect of treating kimchi ingredients (cabbage, hot pepper powder, garlic, ginger, green onion, and leak) with O_3 gas (6 mg/liter/s for 60 min) on the vitamin content, bacterial count and sensory properties of this product were investigated by Kim et al. (1993). The O_3 treatment decreased the microbial counts by 3–4 logs.

The respiratory activity of fresh-cut lettuce was not affected by treatment with ozonated water, despite its strong oxidizing activity (Palou et al., 2001). Zhang et al. (2005) reported that the respiration rate of fresh-cut celery was reduced by ozone treatment. Peach respiration and ethylene production were also not affected after a 3 week exposure at 0.3 ppm ozone (Palou et al., 2005). Removal of ethylene from storage rooms or containers can slow down the ripening process extending post harvest lives of fruits.

The objective of the present study was to determine the effect of ozonation and packaging on physical, chemical, mechanical, and microbiological properties and sensory characteristics of fresh tomatoes stored under refrigeration in an effort to increase product shelf life.

MATERIALS AND METHODS

Sample Treatment

Greenhouse tomatoes were obtained from the Agricultural Cooperative of Preveza, Greece in January, 2009. The samples were transported to the laboratory within 1 hr and washed. Tomatoes were exposed to ozone gas for 1 hr at various concentrations (0.5, 1 and 2 ppm) and stored at 4 ± 1°C for 49 days. The O_3 gas produced by an air tree model C-Lasky L010 ozone generator (Taiwan). An Eco-sensors model OS-4 O_3 detector (Santa Fe, New Mexico USA) was used to measure the concentration of O_3 gas. Half of the samples were placed in commercial crates. The other half were packaged in LDPE bags 35 mm in thickness, having an oxygen permeability of 5700 ml/(m^2 day atm) and a water vapor permeability of 3.6 g/(m^2 day) (T=25°C, RH=0%). Bags were sealed using a Boss N48 thermal sealer (Boss, GmbH, Germany) and kept under refrigeration (4 ± 1°C). A third group of tomatoes (untreated) kept under refrigeration, was taken as the control sample. After 0, 7, 14, 21, 28, 35, and 49 days, randomly chosen tomatoes were collected for analysis.

Moisture Loss of Tomato Fruit

Tomato fruit were selected for uniformity and absence of any visible damage, labeled and weighed before and after exposure to O_3. Weight loss was calculated by the difference in fruit weight on the first day of the experiment and each sampling day. Fruit weight was determined using an electronic balance with two decimal point accuracy.

Firmness of Tomato Fruit

The resistance of tomato to penetration was measured using an Instron model 4411 dynamometer (Instron Ltd., UK). A 6 mm diameter plunger was used at a speed of 20 mm/min. Measurements were made at room temperature (20±1.0°C). Results were expressed as force [in Newtons (N)] required to break the radial pericarp (i.e. skin) of each tomato.

Color Measurement of Tomato Fruit

The color of tomatoes was measured using a Hunter Lab model DP-25L Optical Sensor Colorimeter (Hunter Associates Laboratory, Reston, VA, USA). The colorimeter was calibrated using white and black reference tiles. The color values were expressed as L*(lightness), a* (redness) and b* (yellowness). Tomatoes were sliced 1 cm thick and placed between two glass tiles.

Titratable Acidity (TA) of Tomato Fruit

Tomato pulp samples (20 ml) were placed in a 250 ml erlenmeyer flask and 80 ml distilled water were added. After mixing, the flask content was titrated with 0.1N NaOH using phenolphthalein. Titratable acidity was expressed as g of citric acid per 100 ml of tomato pulp (Redd et al., 1986).

Microbiological Analysis

Microbial counts for mesophilic and psychrotrophic bacteria, yeasts and moulds were monitored. Samples (10 g) were weighed and aseptically transferred to a stomacher bag (Seward Medical, UK), containing 90 ml of Buffered Peptone Water 0.1% (LAB M, UK), and homogenized using a stomacher (Lab Blender 400, Seward Medical) for 30 sec. Appropriate dilutions were prepared. Total viable counts (TVC) and psychrotrophic bacteria were determined using plate count agar (PCA; Merck, Denmark, Germany), after incubation for 2 days at 30°C and 7 days at 7°C respectively. Yeasts and moulds were enumerated using rose Bengal chloroamphenicol agar (RBC; Merck) after incubation for 5 days at 30°C in the dark.

All microbial counts were reported as log CFU/g (colony forming units per gram of sample).

Volatile Compounds Analysis Using Gas Chromatography–Mass Spectrometry (GC-MS)

Prior to analysis, tomato fruit samples were homogenized in a blender for 30 sec. 2 g of each sample, along with 2 ml NaCl (5% w/v) containing internal standard (*n*-amyl alcohol) and a micro-stirring bar were placed in 10 ml glass vials (Schot Duran, Germany) and heated to 45°C in a water bath for 5 min to equilibrate. After equilibration the needle of Solidphase Microextraction (SPME) device was inserted into the vial through the septum and the fiber was exposed to the vial headspace for 15 min. The SPME was performed with a 50/30 μm Divinylbenzene/Carboxen on polydimethylsiloxane. (Supelco, Bellefonte, USA).

The GC/MS analysis was carried out using a Hewlett-Packard HP 6890 gas chromatograph unit combined with a Hewlett-Packard 5973 series mass selective detector. The capillary column used was a DB-5 MS (60 m × 0.320 mm i.d. and 1 μm film thickness, J&W Scientific, Folsom, USA). Helium was used as the carrier gas at a flow rate of 1.2 ml/min. The temperature of injector was 260°C. The oven temperature was programmed at 40°C for 2 min, increased to 120°C at 4°C/min and to 260°C at 10°C/min at which remained for 5 min. The injector was operated in the split mode 2:1. The temperature of ion source and quadrupole was 230°C and 150°C, respectively. The mass range studied was *m/z* 29–400. Identification of volatiles was carried out using the mass spectrum library (Wiley7, Nist 05, J. Wiley & Sons Ltd., West Sussex, UK).

Sensory Evaluation

The samples were evaluated, by a 51 member panel (acceptability test). Panelists (21 females, 30 males) were chosen among graduate students and faculty of the Department of Chemistry, University of Ioannina using the following criteria: Ages between 18 and 60, non smokers, who consume tomatoes on a regular basis. Panelists compared fresh fruit (control) to treated samples and were asked to assess appearance, aroma, taste, and texture using a five point scale (1: "Poor/unsweet/soft" and 5: "excellent/very sweet/firm"). A score of three was taken as the lower limit of acceptability.

Statistical Analysis

Statistical analysis was performed using SPSS (Version 16 for Windows, SPSS, Chicago, IL, USA) and graphs were produced using Microsoft Excel 2007. Data were subjected to one-way analysis of variance (ANOVA). Results are given as mean ± standard deviation of five independent determinations. Significant differences between mean values were determined using the least significant difference LSD ($p < 0.05$).

DISCUSSION AND RESULTS

Effect of Ozonation on Moisture Loss in Tomato Fruit

Storage time, LDPE packaging and ozonation were found to significantly affect ($p < 0.05$) the percentage weight loss of samples (Figure 2). After 49 days of storage, tomatoes packed in LDPE, regardless of ozonation dose, showed the lowest percentage of weight loss. This is due to the excellent barrier properties of LDPE to moisture. Weight loss in ozonated samples in crates was significantly affected ($p < 0.05$) after 7 days while in samples packed in LDPE after day 0 of storage. The lowest percentage weight loss was recorded in samples 2 ppm O_3-LDPE and the highest in control samples. This is due to the uncontrolled ripening in control fruits, as ripening in tomatoes is climacteric which shows a sudden increase in ethylene production and thus an increase in respiration rate (Anonymous, 2004). This higher respiration rate also resulted in higher transpiration of water from the fruit surface which led to an increase in weight loss (Ball, 1997).

The Packaging reduced the weight loss of tomato fruit by 40–50% as compared to control samples throughout storage (Sammi and Masud, 2007). These results are in agreement with those of Batu and Thompson (1998) who reported that pack sealed tomatoes showed less weight loss as compared to controls after 60 days of storage. Lower weight loss was due to the barrier of the packaging material to primarily moisture and secondarily to oxygen creating a passive modified atmosphere in the immediate environment of the fruit.

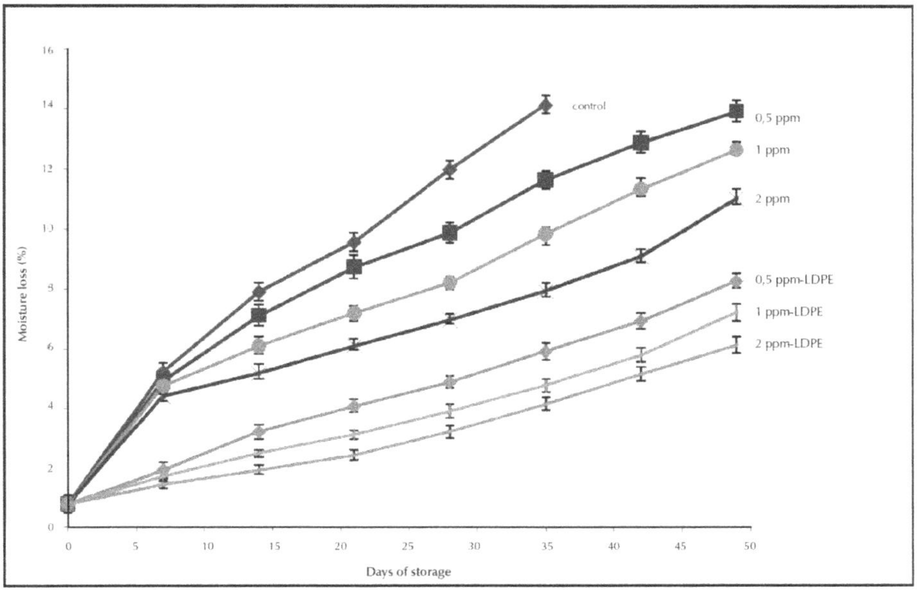

Figure 2. Effect of ozonation, packaging and storage time on moisture loss (%) in tomatoes at 4°C.

Effect of Ozonation on Tomato Firmness

Results regarding the resistance of tomato fruit to penetration are shown in Table 1. Changes in this parameter were statistically significant ($p < 0.05$) after day 7 for control, ozonated at 0.5 ppm, 0.5 ppm-LDPE, and 1 ppm samples; after day 14 for 1 ppm-LDPE samples and day 21 for 2 ppm and 2 ppm-LDPE samples. Ozonation dose significantly affected ($p < 0.05$) resistance to penetration after day 7 for all samples.

Texture played an important role in acceptability of the tomato fruit being determined by the fruit morphological and physiological characteristics: Epicarp firmness, amount of locule tissue and maturity stage (Chiesa et al., 1998). Tzortzakis et al. (2007) reported that tomatoes stored for 6 days in an ozone-enriched atmosphere retained their texture (*Lycopersicon esculentum* L. cv. Carousel).

Aguayo et al. (2006) reported that the firmness of tomato slices [stored up to 15 days at 5°C under air (control) or under a cyclic flow of O_3 (4±0.5 µl/l of O_3 for 30 min every 3 hr)] did not change after O_3 exposure whereas a significant effect was shown with storage time. In sliced tomatoes, the reduction in firmness was ca. 9.2%, 16%, and 35% after 5, 12, and 15 days of storage, respectively. However, in whole tomatoes O_3 treatment reduced softness. After 15 days of storage, the reduction in firmness was 21% after O_3 treatment, and 28% for control fruit.

Table 1. Effect of ozonation, packaging and storage time on tomato fruit resistance to penetration.

DAYS	LOAD (N)						
	control	0,5 ppm	0,5 ppm-LDPE	1 ppm	1 ppm-LDPE	2 ppm	2 ppm-LDPE
0	16.27±0.87	16.27±0.87	16.27±0.87	16.27±0.87	16.27±0.87	16.27±0.87	16.27±0.87
7	14.59±0.86	14.70±0.83	14.94±0.79	15.23±0.84	15.54±0.80	15.77±0.81	16.19±0.82
14	11.14±0.42	12.01±0.43	12.91±0.40	13.80±0.45	14.65±0.37	15.41±0.32	16.08±0.33
21	9.64±0.29	11.75±0.26	12.54±0.27	13.28±0.28	14.16±0.26	14.79±0.28	15.44±0.30
28	8.96±0.32	11.13±0.28	11.86±0.25	12.88±0.21	13.58±0.28	14.12±0.21	14.83±0.25
35	7.32±0.31	9.93±0.26	10.69±0.27	12.37±0.22	12.99±0.27	13.61±0.25	14.14±0.22
42	–	9.36±0.24	10.38±0.26	11.95±0.28	12.48±0.21	12.98±0.23	13.52±0.25
49	–	8.88±0.21	9.42±0.24	10.79±0.26	11.32±0.23	11.83±0.22	12.77±0.29

Effect of Ozonation on Tomato Color

Results regarding color measurement of tomato fruit are shown in Table 2. Changes in L* values were significantly affected ($p < 0.05$) by storage time. In contrast, ozonation dose did not significantly affect ($p > 0.05$) L* values. Furthermore, there were no statistically significant changes in a* and b* values ($p > 0.05$) either with storage time or ozonation dose.

Color is the primary factor considered by consumers in quality assessment of fruits and vegetables (Gejima et al., 2003). Consequently, a bright red color constitutes a significant quality parameter for consumer acceptance of tomato products (Sharma et al., 1996). Colorimeters express colors in terms of tristimulus parameters L* (white to black), a* (green to red) and b* (blue to yellow) respectively (López Camelo et al., 2004). The lycopene content during ripening of tomatoes is considered the main factor responsible for color development (Meredith and Purcell, 1996; Thomson et al., 1965). The change in color of tomato fruit during ripening is a function of time and storage temperature (Tijskens and Evelo, 1994).

Tiwari et al. (2009) reported that L*, a* and b* color values of tomato juice samples were significantly affected by O_3 concentration and treatment time. After ozonation juice samples were observed to be lighter in color that is increased L* value, whereas a* and b* values were found to decrease with increase in treatment time and O_3 concentration.

Ozonation of tomato slices (4±0.5 μl/l for 30 min every 3 hr) did not affect color for 5 days. However, when samples were stored up to 15 days an increase in L* parameter compared to the initial value after cutting was observed (Aguayo et al., 2006). This was probably due to the senescence symptom accelerated by the cut (Mencarelli and Saltveit, 1988). Furthermore, Maguire and Solverg (1980) found that ozonation induced lycopene accumulation in tomatoes. Nevertheless, Perez et al. (1999) measured lower anthocyanin content in strawberry and Singh et al. (2002) observed discoloration in lettuce after ozone-treatment. Likewise, Skog and Chu (2000) reported browning occurring on the cut surface in broccoli after ozonation.

Effect of Ozonation on Titratable Acidity (Ta) of Tomatoes

Results for titratable acidity of tomato pulp are shown in Figure 3. Both storage time and O_3 concentration were found to have a statistically significant effect ($p < 0.05$) on citric acid concentration for all samples after day 7.

According to Bhattacharya et al. (2004) acidity is often used as an index of maturity, as acid decreases during ripening of fruit. Salvador et al. (2006) stated that during ripening acidity declines after the first appearance of yellow color. Reducing in concentration of malic and citric acid during the ripening process may be the main factor responsible for the reduction in titratable acidity during storage (Stevens, 1972).

Table 2. Effect of ozonation, packaging and storage time of tomato color.

DAYS	0	7	14	21	28	35	42	49
Control								
L	31.96±2.92	37.09±3.48	38.39±3.33	40.34±3.49	41.51±3.54	43.01±3.52	-	-
a	12.56±1.68	12.48±1.69	11.86±1.73	11.74±1.72	11.70±1.75	11.65±1.82	-	-
b	12.53±1.64	12.35±1.65	11.31±1.76	11.26±1.74	11.23±1.78	11.17±1.81	-	-
0,5 ppm O_3								
L	31.96±2.92	38.03±3.59	39.18±3.23	41.84±3.35	42.61±3.46	43.22±3.59	44.39±3.65	44.71±3.91
a	12.56±1.68	12.42±1.86	11.72±1.89	11.68±1.91	11.64±1.94	11.61±1.88	11.59±1.81	11.51±1.92
b	12.53±1.64	12.30±1.75	11.26±1.82	11.23±1.87	11.20±1.91	11.15±1.94	11.06±1.99	10.99±1.93
0,5 ppm O_3-LDPE								
L	31.96±2.92	38.43±3.47	39.99±3.48	42.28±3.19	43.33±3.39	43.52±3.33	44.61±3.35	45.38±3.81
a	12.56±1.68	12.39±1.67	11.66±1.68	11.61±1.71	11.59±1.76	11.55±1.74	11.51±1.69	11.31±1.66
b	12.53±1.64	12.29±1.67	11.24±1.69	11.20±1.74	11.16±1.73	11.08±1.81	11.00±1.75	10.93±1.78
1 ppm O_3								
L	31.96±2.92	38.81±3.58	40.69±3.24	42.86±3.37	43.81±3.43	43.99±3.67	44.72±3.61	45.84±3.65
a	12.56±1.68	12.34±1.72	11.50±1.75	11.21±1.76	11.13±1.80	11.04±1.89	10.90±1.93	10.82±1.74
b	12.53±1.64	11.98±1.68	11.21±1.79	11.11±1.73	11.04±1.77	10.80±1.80	10.78±1.75	10.74±1.74
1 ppm O_3-LDPE								
L	31.96±2.92	39.18±3.26	41.32±3.21	43.13±3.46	44.21±3.31	44.81±3.86	45.31±3.47	46.32±3.97
a	12.56±1.68	12.01±1.79	11.13±1.72	10.92±1.71	10.84±1.68	10.71±1.72	10.69±1.71	10.66±1.69
b	12.53±1.64	11.90±1.78	11.10±1.75	10.81±1.72	10.77±1.71	10.63±1.88	10.51±1.79	10.40±1.73
2 ppm O_3								
L	31.96±2.92	39.43±3.62	42.08±3.73	44.27±3.41	45.32±3.59	45.56±3.79	45.61±3.88	46.75±3.96
a	12.56±1.68	11.91±1.84	11.06±1.86	10.79±1.89	10.71±1.90	10.62±1.92	10.57±1.98	10.54±1.93
b	12.53±1.64	11.84±1.84	10.85±1.92	10.67±1.95	10.61±1.89	10.54±1.83	10.43±1.94	10.21±1.99
2 ppm O_3-LDPE								
L	31.96±2.92	40.37±3.52	42.88±3.29	45.22±3.43	45.81±3.57	46.11±3.61	46.47±3.75	47.21±3.79
a	12.56±1.68	11.87±1.71	10.74±1.76	10.69±1.52	10.54±1.89	10.43±1.96	10.39±1.89	10.32±1.81
b	12.53±1.64	11.78±1.87	10.67±1.74	10.53±1.83	10.42±1.90	10.37±1.91	10.09±1.93	9.90±1.95

Aguayo et al. (2006) reported that TA of tomato juice ranged from 3.5 to 4.0 g citric acid/l and between 3.6 and 3.7 g citric acid/l in whole tomatoes. Neither storage time (15 days) nor O_3 treatment (4±0.5 µl/l for 30 min every 3 hr) had a significant effect on this parameter. At the end of storage, whole tomatoes kept in air and O_3 had higher citric acid content than slices, probably due to organic acid oxidation in the cut fruit.

Tzortzakis et al. (2007) reported that citric acid concentration declined during ripening after ozone-treatment (0.05–1.0 µmol/mol for 12 days). Ozonation resulted in insignificant changes in citrate or malate levels, but citrate levels declined more rapidly in fruit previously subjected to low-level ozonation (0.05 µmol/mol) than in control samples. Shalluf et al. (2007) reported that ozone-treatment (2, 7, and 21 mg O_3/g tomato for 20 min, 1 hr and 3 hr respectively) had no significant effect of total acidity of tomatoes after 14 days. Similarly, no significant effect of O_3 on TA was reported by Garcia et al. (1998) after storage of three varieties of oranges. Perez et al (1999) reported a similar finding for storage of strawberries.

Tiwari et al. (2009) observed no statistically significant changes ($p > 0.05$) in TA of ozonated tomato samples (1.6–7.8% w/w for 0–10 min). Similar results were reported for ozonated orange juice (Tiwari et al., 2008) and apple cider (Choi and Nielsen, 2005).

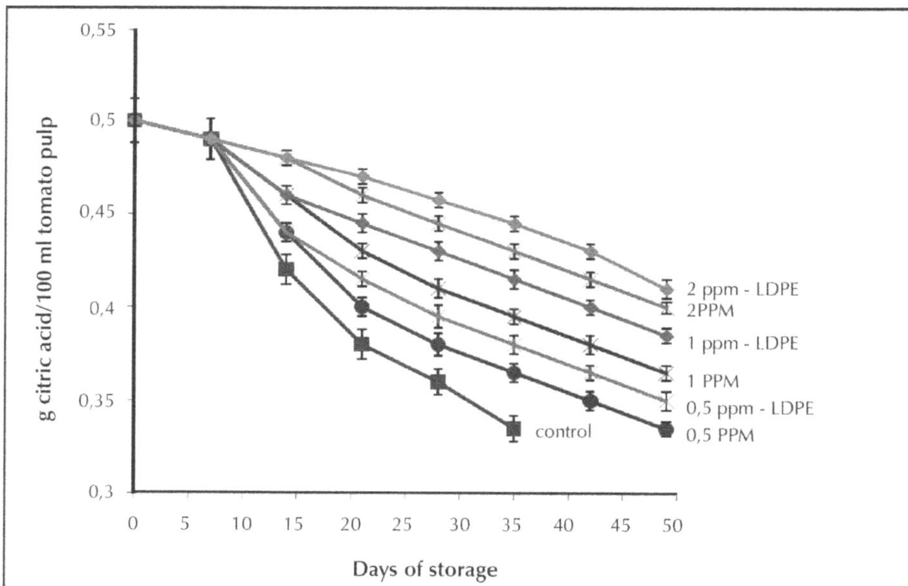

Figure 3. Effect of ozonation, packaging and storage time on Titratable Acidity of tomato pulp at 4°C.

Microbiological Analysis

The effect of ozonation, packaging and storage time on TVC, psychrotrophs and yeasts and moulds of tomato fruit is shown in Figures 4, 5, and 6. The TVC, psychrotrophs as well as yeasts and moulds were affected ($p < 0.05$) by ozonation after day 7. Storage time significant affected ($p < 0.05$) TVC between 0 and 28 day, psychrotrophs between 0 and 35 day and yeasts/moulds between 0 and 49 day for ozonated samples. Storage time significant affected ($p < 0.05$) TVC, psychrotrophs and yeasts/moulds between 0 and 35 day for control samples.

Aguayo et al. (2006) reported that the ozonation (dose: 4±0.5 μl/l) reduced TVC as well as fungal counts, in agreement with reports of Baranovskaya et al. (1979). On day 6, the reduction in TVC and psychrotrophs was 0.38 and 0.56 log cycles respectively. Reduction in microbial populations was higher toward the end of storage. No fungal growth was detected in the O_3-treated tomato slices after 15 days of storage. A continuous exposure to O_3 (0.3 μl/l) delayed green and blue mould growth in citrus (Palou et al., 2001) and decay in blackberries (Barth et al., 1995). On the other hand Rice et al. (1982) found that yeasts are more readily attacked by O_3 than bacteria.

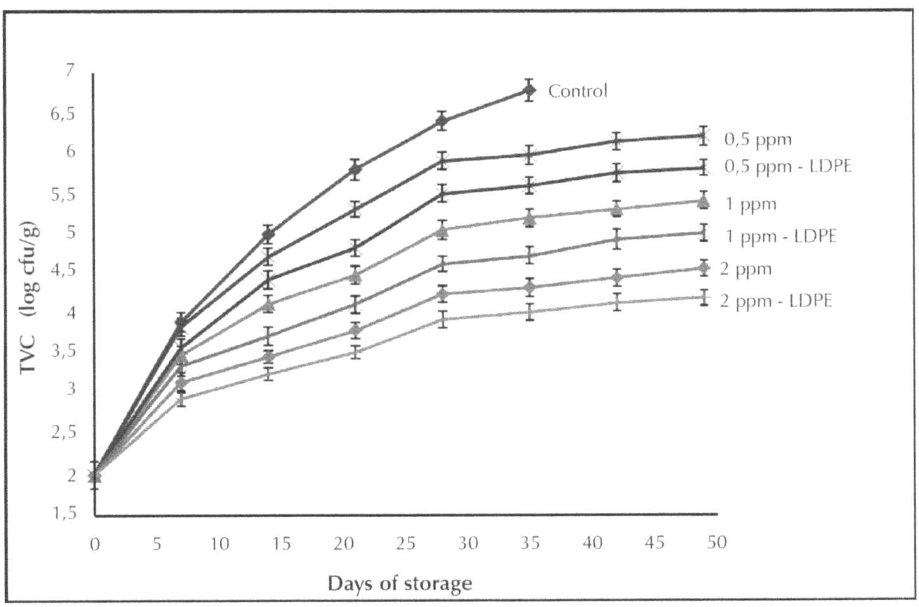

Figure 4. Effect of ozonation, packaging and storage time on total viable count in tomatoes.

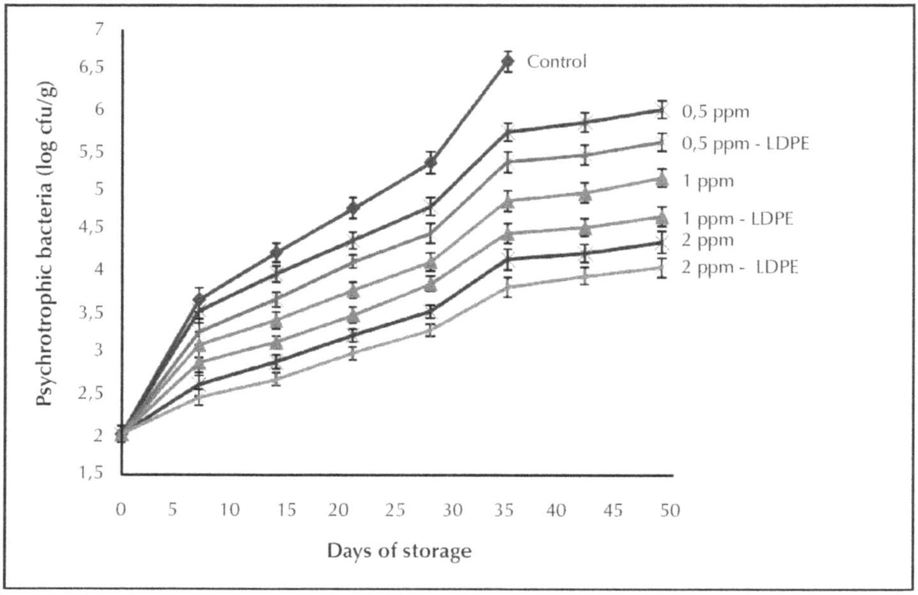

Figure 5. Effect of ozonation, packaging and storage time on psychrotrophic count in tomatoes.

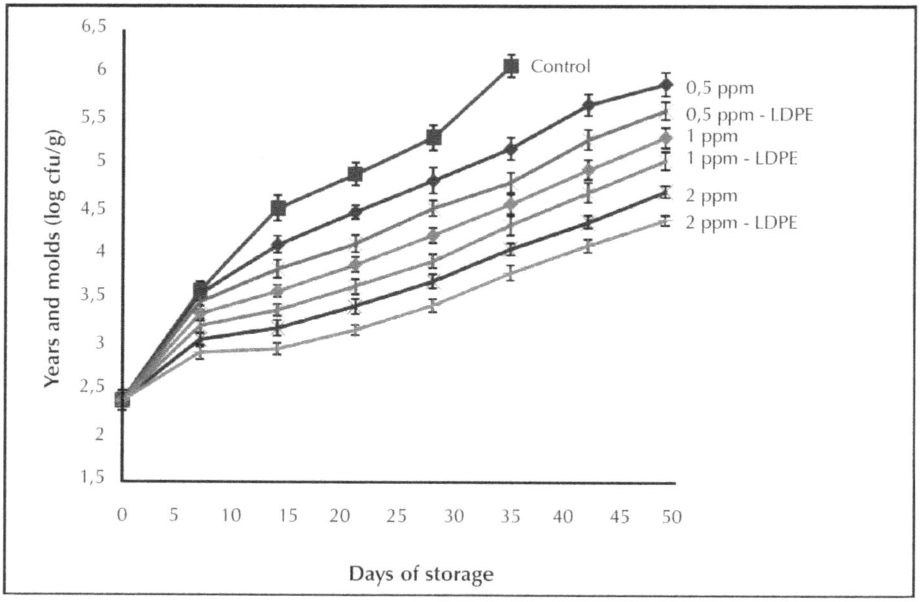

Figure 6. Effect of ozonation, packaging and storage time on yeasts and moulds growth in tomatoes.

GC-MS Volatile Compounds Analysis

Volatile components of tomato fruit (mg/kg) for all treatments are shown in Tables 4–10. A typical chromatogram of volatile components of tomato fruit is shown in Figure 7. Results showed that there were no statistically significant changes ($p > 0.05$) recorded in most volatile-aromatic components of fresh tomato samples either with storage time or dose of ozonation. An exception to this rule was ethanol which decreased with ozonation and increased with storage time. The Packaging in LDPE bags contributes to retention of ethanol. 1-penten-3-one, trans 2-pentenal, n-hexanal, trans 2-heptenal and geranyl acetone increased while cis 3-hexenol decreased in concentration with storage time.

Characteristic tomato flavor is the result of interaction between volatiles and taste components. Of the ca. 400 volatile compounds determined in tomatoes, only a limited number are considered essential to tomato aroma. Volatiles in fresh tomatoes are formed from lipids, carotenoids, amino acids, terpenoids (C_{10} and C_{15}), lignin and other sources (Baldwin et al., 1991; Buttery et al., 1987; Petro-Turza, 1987). Although flavor perception has an important odor component, several investigators reported that sugars and organic acids contribute more to tomato flavor than volatile compounds. According to Baldwin et al. (1991) many volatiles increased during ripening along with the increased synthesis of lycopene, the pigment responsible for red color in tomatoes.

Aldehydes contribute significantly to the aroma of fresh tomatoes. Analysis of tomato samples identified 15 aldehydes: Butanal, pentanal, trans-2-pentenal, n-hexanal, heptenal, trans-2-heptenal, cis-4-heptenal, 2-octenal, nonanal, decanal, benzaldehyde, geranial, cis-4-decenal, trans-trans-2,4-decadienal and beta-cyclocitral.

The n-hexanal is important in fresh tomato flavor (Petro-Turza, 1987). It is produced by linoleic acid 13 hydroperoxides and contributes to fruit aroma with musty and moldy tones (Alonso et al., 2009). According to Buttery et al. (1987) n-hexanal is considered the largest contributor to the aroma of tomatoes. It increased with storage time, which is consistent with Buttery and Ling, (1993). Nonanal and trans 2-heptenal derive from lipids and specifically from unsaturated fatty acids. These compounds are considered important in the flavor of tomato (Baldwin et al., 1991). Trans 2-heptenal increased in all samples with time (Buttery and Ling, 1993). Trans 2-pentenal occurred in small amounts in the aroma of tomatoes, but may affect product overall flavor (Kazeniac and Hall, 1970). This aldehyde increased in all samples with storage time.

Table 4. Volatile components of control tomato samples (mg/kg).

VOLATILE COMPOUND	RT (min)	7th DAY	21th DAY	35th DAY
Ethanol	2.33	0.53	1.72	2.53
Butanal	2.83	0.08	0.09	0.10
Acetic acid, ethyl ester	4.73	0.14	0.16	0.15
1-penten-3-one	7.36	0.23	0.36	0.50

Table 4. *(Continued)*

VOLATILE COMPOUND	RT (min)	7th DAY	21th DAY	35th DAY
Pentanal	7.83	0.25	0.23	0.21
3-methyl-1-butanol	9.40	0.27	0.25	0.26
Trans-2-pentenal	10.27	0.10	0.15	0.18
IS. 1-pentanol	10.80	20.25	20.25	20.25
N-Hexanal	12.17	2.21	3.11	3.87
Cis-3-hexenol	14.54	1.33	0.83	0.36
1-Hexanol	14.94	0.64	0.66	0.65
Cis-4-Heptenal	16.22	0.05	0.06	0.07
N-Heptanal	16.31	0.05	0.06	0.04
Acetic acid, pentyl ester	16.68	0.73	0.69	0.70
Hexanoic acid, methyl ester	17.11	0.05	0.05	0.03
2-methyl-3-Heptanone	17.61	0.13	0.11	0.15
Trans-2-Heptenal	18.45	0,12	0.23	0.34
Benzaldehyde	18.82	0.14	0.13	0.15
2, 6-dimethyl-4-Heptanone	18.91	0.19	0.17	0.18
6-methyl-5-hepten-2-one	19.42	1.29	1.28	1.27
Furan, 2-pentyl	19.67	0.14	0.12	0.13
Propanoic acid, pentyl ester	20.07	0.89	0.84	0.85
Hexyl acetate	20.42	0.06	0.07	0.08
Butanoic acid, 3-methyl, -butyl	21.38	0.55	0.52	0.54
1, 2-dichloro-benzene	21.48	0.09	0.08	0.11
2-Octenal	22.08	0.32	0.34	0.33
N-Octanol	22.40	0.12	0.14	0.15
Butanoic acid, pentyl ester	23.14	0.21	0.17	0.19
Nonanal	23.62	0.24	0.27	0.25
Carvacrol	23.86	0.04	0.03	0.05
Cis-4-Decenal	26.39	0.09	0.10	0.08
Decanal	26.69	0.08	0.05	0.05
β-Cyclocitral	27.28	0.07	0.05	006
Geranial	28.17	0.08	0.09	0.10
Trans, trans 2,4-Decadienal	28.75	0.08	0.09	0.08
Eugenol	29.94	0.05	0.04	0.06
Geranyl acetone	31.42	0.36	0.43	0.52
β-ionone	32.08	0.08	0.07	0.09

Table 5. Volatile components of tomato (mg/kg) after exposure to 0.5 ppm of ozone gas.

VOLATILE COMPOUND	RT (min)	7th DAY	21th DAY	35th DAY	49th DAY
Ethanol	2.33	0.50	0.68	1.10	1.22
Butanal	2.83	0.07	0.06	0.08	0.07
Acetic acid, ethyl ester	4.73	0.14	0.18	0.17	0.19
1-penten-3-one	7.36	0.25	0.35	0.51	0.68
Pentanal	7.83	0.26	0.25	0.22	0.25
3-methyl-1-butanol	9.40	0.24	0.22	0.23	0.25
trans 2-pentenal	10.27	0.11	0.15	0.17	0.19
IS. 1-pentanol	10.80	20.25	20.25	20.25	20.25
N-Hexanal	12.17	2.29	2.62	3.10	3.83
Cis-3-hexenol	14.54	1.25	0.75	0.58	0.30
1-Hexanol	14.94	0.66	0.63	0.65	0.64
Cis-4-Heptenal	16.22	0.08	0.07	0.07	0.06
N-Heptanal	16.31	0.07	0.07	0.05	0.05
Acetic acid, pentyl ester	16.68	0.59	0.65	0.63	0.64
Hexanoic acid, methyl ester	17.11	0.03	0.06	0.03	0.05
2-methyl-3-Heptanone	17.61	0.12	0.09	0.10	0.11
Trans-2-Heptenal	18.45	0.15	0.25	0.32	0.37
Benzaldehyde	18.82	0.10	0.14	0.12	0.13
2,6-dimethyl-4-Heptanone	18.91	0.15	0.18	0.15	0.16
6-methyl-5-hepten-2-one	19.42	1.27	1.28	1.24	1.29
Furan, 2-pentyl	19.67	0.16	0.12	0.13	0.14
Propanoic acid, pentyl ester	20.07	0.74	0.76	0.72	0.73
Hexyl acetate	20.42	0.06	0.06	0.08	0.07
Butanoic acid, 3-methyl, -butyl	21.38	0.55	0.51	0.50	0.53
1, 2-dichloro-benzene	21.48	0.09	0.09	0.10	0.11
2-Octenal	22.08	0.34	0.30	0.32	0.31
N-Octanol	22.40	0.13	0.12	0.11	0.12
Butanoic acid, pentyl ester	23.14	0.22	0.21	0.20	0.19
Nonanal	23.62	0.22	0.22	0.20	0.18
Carvacrol	23.86	0.08	0.05	0.04	0.04
Cis-4-Decenal	26.39	0.10	0.16	0.15	0.13
Decanal	26.69	0.04	0.09	0.06	0.06
β-Cyclocitral	27.28	0.08	0.09	0.07	0.09
Geranial	28.17	0.08	0.09	0.09	0.10
Trans, trans 2, 4-Decadienal	28.75	0.05	0.06	0.06	0.07
Eugenol	29.94	0.04	0.05	0.04	0.05
Geranyl acetone	31.42	0.34	0.46	0.53	0.59
β-ionone	32.08	0.07	0.05	0.03	0.04

Alcohols found in the samples were: Ethanol, 3-methyl butanol, 1-hexanol, cis 3-hexenol and n-octanol, eugenol and carvacrol. Cis 3-hexenol is a volatile compound that is mainly found in fresh tomatoes and juice (Markovic et al., 2007) with pleasant fragrance (Butttery et al., 1987). High amounts of ethanol were reported by Katayama et al. (1967) in fresh tomato juice. Possibly, ethanol is produced from the reduction of large amounts of acetaldehyde (Eriksson, 1968). Nelson and Hoff (1969) reported that acetaldehyde decreased while ethanol sharply increased during storage. Eugenol and carvacrol have been found in small quantities in several studies involving tomato processing. Nevertheless they contribute to tomato aroma with characteristic fruity notes (Buttery and Ling, 1993; Koukol and Crom, 1961; Markovic et al., 2007).

Table 6. Volatile components of tomato (mg/kg) after exposure to 1 ppm of ozone gas.

VOLATILE COMPOUND	RT (min)	7th DAY	21th DAY	35th DAY	49th DAY
Ethanol	2.33	0.42	0.58	0.69	1.01
Butanal	2.83	0.08	0.07	0.06	0.08
Acetic acid, ethyl ester	4.73	0.14	0.15	0.15	0.16
1-penten-3-one	7.36	0.26	0.46	0.57	0.69
Pentanal	7.83	0.22	0.23	0.21	0.25
3-methyl-1-butanol	9.40	0.24	0.24	0.26	0.25
Trans-2-pentenal	10.27	0.13	0.16	0.18	0.21
IS. 1-pentanol	10.80	20.25	20.25	20.25	20.25
N-Hexanal	12.17	2.23	2.92	3.32	3.74
Cis-3-hexenol	14.54	1.29	0.71	0.47	0.31
1-Hexanol	14.94	0.64	0.66	0.67	0.65
Cis-4-Heptenal	16.22	0.05	0.06	0.04	0.05
N-Heptanal	16.31	0.06	0.07	0.08	0.05
Acetic acid, pentyl ester	16.68	0.68	0.69	0.73	0.70
Hexanoic acid, methyl ester	17.11	0.03	0.03	0.04	0.06
2-methyl-3-Heptanone	17.61	0.12	0.09	0.10	0.12
Trans-2-Heptenal	18.45	0.15	0.21	0.32	0.39
Benzaldehyde	18.82	0.16	0.16	0.15	0.14
2, 6-dimethyl-4-Heptanone	18.91	0.14	0.16	0.17	0.15
6-methyl-5-hepten-2-one	19.42	1.24	1.25	1.27	1.28
Furan, 2-pentyl	19.67	0.19	0.17	0.18	0.19
Propanoic acid, pentyl ester	20.07	0.69	0.70	073	0.72
Hexyl acetate	20.42	0.09	0.08	0.07	0.07
Butanoic acid, 3-methyl, -butyl	21.38	0.58	0.55	0.56	0.57
1, 2-dichloro-benzene	21.48	0.09	0.10	0.11	0.12
2-Octenal	22.08	0.31	0.30	0.29	0.31

Table 6. *(Continued)*

VOLATILE COMPOUND	RT (min)	7th DAY	21th DAY	35th DAY	49th DAY
N-Octanol	22.40	0.12	0.12	0.13	0.14
Butanoic acid, pentyl ester	23.14	0.21	0.22	0.22	0.19
Nonanal	23.62	0.22	0.21	0.25	0.24
Carvacrol	23.86	0.07	0.06	0.06	0.06
Cis-4-Decenal	26.39	0.07	0.09	0.11	0.13
Decanal	26.69	0.04	0.03	0.06	0.03
β-Cyclocitral	27.28	0.06	0.04	0.10	0.08
Geranial	28.17	0.12	0.09	0.09	0.10
Trans, trans 2, 4-Decadienal	28.75	0.06	0.05	0.05	0.05
Eugenol	29.94	0.05	0.04	0.06	0.10
Geranyl acetone	31.42	0.31	0.39	0.44	0.58
β-ionone	32.08	0.08	0.05	0.05	0.06

Ketones identified in tomato samples were: 2-methyl-3-heptanone, 2, 6-dimethyl-4-heptanone, 6-methyl-5-hepten-2-one, 1-pentene-3-one, geranyl acetone, and beta-ionone. Ketones generally contribute to the fruity aroma of tomatoes (Heath, 1978). Reducing the levels of lycopene affects the production of important flavor compounds belonging to ketones (Buttery and Ling, 1993). The 6-Methyl-5-hepten-2-one and 1-penten-3-one were reported to be important for tomato aroma (Buttery et al., 1971, 1987, 1988; Ho and Ichimura, 1982; Petro-Turzak, 1987). The 6-Methyl-5-epten-2-one is related to carotenoids and is characterized by a floral aroma (Baldwin et al., 1991). Beta ionone is responsible for the sweet and fruity aroma of tomatoes (Kazeniac and Hall, 1970; Tandon, 1997). It is produced by oxidative degradation of beta-carotene (Cole et al., 1957).

Table 7. Volatile components of tomato (mg/kg) after exposure to 2 ppm of ozone gas.

VOLATILE COMPOUND	RT (min)	7th DAY	21th DAY	35th DAY	49th DAY
Ethanol	2.33	0.44	0.52	0.73	0.91
Butanal	2.83	0.08	0.08	0.09	0.07
Acetic acid, ethyl ester	4.73	0.13	0.14	0.13	0.16
1-penten-3-one	7.36	0.27	0.41	0.59	0.70
Pentanal	7.83	0.28	0.26	0.26	0.27
3-methyl-1-butanol	9.40	0.27	0.22	0.25	0.23
Trans-2-pentenal	10.27	0.13	0.15	0.18	0,19
IS. 1-pentanol	10.80	20.25	20.25	20.25	20.25
N-Hexanal	12.17	2.32	2.75	3.21	3.81

Table 7. *(Continued)*

VOLATILE COMPOUND	RT (min)	7th DAY	21th DAY	35th DAY	49th DAY
Cis-3-hexenol	14.54	1.28	0.69	0.49	0.35
1-Hexanol	14.94	0.63	0.68	0.68	0.67
Cis 4-Heptenal	16.22	0.04	0.06	0.07	0.06
N-Heptanal	16.31	0.07	0.04	0.05	0.06
Acetic acid, pentyl ester	16.68	0.66	0.68	0.73	0.71
Hexanoic acid, methyl ester	17.11	0.06	0.05	0.04	0.05
2-methyl-3-Heptanone	17.61	0.15	0.10	0.12	0.13
Trans-2-Heptenal	18.45	0.16	0.27	0.34	0.42
Benzaldehyde	18.82	0.14	0.15	0.13	0.16
2,6-dimethyl-4-Heptanone	18.91	0.13	0.15	0.16	0.15
6-methyl-5-hepten-2-one	19.42	1.24	1.21	1.25	1.23
Furan, 2-pentyl	19.67	0.19	0.16	0.17	0.18
Propanoic acid, pentyl ester	20.07	0.74	0.78	0.75	0.70
Hexyl acetate	20.42	0.06	0.07	0.07	0.08
Butanoic acid, 3-methyl, -butyl	21.38	0.58	0.55	0.57	0.58
1,2-dichloro-benzene	21.48	0.10	0.12	0.09	0.08
2-Octenal	22.08	0.35	0.33	0.34	0.32
N-Octanol	22.40	0.13	0.15	0.12	0.16
Butanoic acid, pentyl ester	23.14	0.27	0.24	0.25	0.27
Nonanal	23.62	0.18	0.26	0.19	0.25
Carvacrol	23.86	0.05	0.04	0.08	0.07
Cis-4-Decenal	26.39	0.12	0.10	0.09	0.09
Decanal	26.69	0.07	0.05	0.09	0.08
β-Cyclocitral	27.28	0.10	0.07	0.09	0.08
Geranial	28.17	0.10	0.11	0.12	0.13
Trans, trans 2,4-Decadienal	28.75	0.04	0.07	0.07	0.09
Eugenol	29.94	0.04	0.07	0.08	0.04
Geranyl acetone	31.42	0.37	0.44	0.50	0.57
β-ionone	32.08	0.05	0.06	0.07	0.07

The following six esters were identified: Acetic acid ethyl ester, acetic acid pentyl ester, hexanoic acid methyl ester, propanoic acid pentyl ester, hexyl acetate, and butanoic acid pentyl ester. Their concentration did not change with storage time and ozonation dose in all samples. Many esters have been reported in numerous studies but their importance to tomato aroma has not been documented (Dalal et al., 1967; Shah et al., 1969). Esters that give sweet fruity aroma include acetic acid pentyl ester, hexanoic acid methyl ester, (Schormuller Kochmann, 1969) and hexyl acetate (Markovic et al., 2007).

Furans identified include: 3-methyl furan and 2-pentyl furan. Furans are heterocyclic compounds containing oxygen, associated with tomato ripening giving buttery fragrances. Their amount did not change with storage time (Buttery et al., 1971). A unique aromatic hydrocarbon found in tomato samples was 1,2-dichloro benzene. Several derivatives of benzene have been reported among the aromatic constituents of tomatoes and their concentration did not change with storage time (Servili et al., 2000).

Table 8. Volatile components of tomato (mg/kg) after exposure to 0.5 ppm of ozone gas packed in LDPE bags.

VOLATILE COMPOUND	RT (min)	7th DAY	21th DAY	35th DAY	49th DAY
Ethanol	2.33	0.40	0.64	0.85	1.20
Butanal	2.83	0.09	0.08	0.07	0.08
Acetic acid, ethyl ester	4.73	0.13	0.15	0.14	0.16
1-penten-3-one	7.36	0.26	0.38	0.50	0.66
Pentanal	7.83	0.28	0.26	0.23	0.25
3-methyl-1-butanol	9.40	0.26	0.26	0.25	0.25
Trans-2-pentenal	10.27	0.14	0.17	0.19	0.19
IS. 1-pentanol	10.80	20.25	20.25	20.25	20.25
N-Hexanal	12.17	2.37	2.75	3.16	3.60
Cis-3-hexenol	14.54	1.27	0.73	0.54	0.33
1-Hexanol	14.94	0.66	0.67	0.65	0.66
Cis-4-Heptenal	16.22	0.07	0.06	0.07	0.05
N-Heptanal	16.31	0.07	0.05	0.06	0.04
Acetic acid, pentyl ester	16.68	0.68	0.65	0.72	0.70
Hexanoic acid, methyl ester	17.11	0.05	0.04	0.06	0.07
2-methyl-3-Heptanone	17.61	0.11	0.14	0.15	0.12
Trans-2-Heptenal	18.45	0.18	0.29	0.34	0.45
Benzaldehyde	18.82	0.14	0.15	0.16	0.17
2,6-dimethyl-4-Heptanone	18.91	0.19	0.17	0.18	0.17
6-methyl-5-hepten-2-one	19.42	1.23	1.27	1.22	1.25
Furan, 2-pentyl	19.67	0.16	0.18	0.17	0.16
Propanoic acid, pentyl ester	20.07	0.64	0.66	0.61	0.65
Hexyl acetate	20.42	0.07	0.08	0.07	0.06
Butanoic acid, 3-methyl, -butyl	21.38	0.53	0.54	0.55	0.56
1,2-dichloro-benzene	21.48	0.09	0.10	0.08	0.09
2-Octenal	22.08	0.34	0.35	0.36	0.35
N-Octanol	22.40	0.13	0.14	0.15	0.12
Butanoic acid, pentyl ester	23.14	0.24	0.21	0.22	0.19

Table 8. *(Continued)*

VOLATILE COMPOUND	RT (min)	7th DAY	21th DAY	35th DAY	49th DAY
Nonanal	23.62	0.21	0.17	0.18	0.19
Carvacrol	23.86	0.08	0.06	0.07	0.09
Cis-4-Decenal	26.39	0.06	0.07	0.06	0.05
Decanal	26.69	0.09	0.08	0.07	0.06
β-Cyclocitral	27.28	0.05	0.06	0.05	0.04
Geranial	28.17	0.07	0.08	0.08	0.10
Trans, trans 2,4-Decadienal	28.75	0.07	0.06	0.05	0.06
Eugenol	29.94	0.09	0.09	0.12	0.13
Geranyl acetone	31.42	0.32	0.40	0.47	0.59
β-ionone	32.08	0.04	0.06	0.05	0.07

Table 9. Volatile components of tomato (mg/kg) after exposure to 1 ppm of ozone gas packed in LDPE bags.

VOLATILE COMPOUND	RT (min)	7th DAY	21th DAY	35th DAY	49th DAY
Ethanol	2.33	0.41	0.53	0.67	0.99
Butanal	2.83	0.09	0.09	0.08	0.10
Acetic acid, ethyl ester	4.73	0.14	0.13	0.14	0.15
1-penten-3-one	7.36	0.28	0.41	0.55	0.69
Pentanal	7.83	0.24	0.22	0.21	0.23
3-methyl-1-butanol	9.40	0.22	0.21	0.20	0.21
Trans-2-pentenal	10.27	0.13	0.15	0.18	0.20
IS. 1-pentanol	10.80	20.25	20.25	20.25	20.25
N-Hexanal	12.17	2.45	2.94	3.14	3.59
Cis-3-hexenol	14.54	1.27	0.78	0.52	0.32
1-Hexanol	14.94	0.62	0.64	0.65	0.64
Cis-4-Heptenal	16.22	0.04	0.05	0.06	0.07
N-Heptanal	16.31	0.09	0.08	0.05	0.07
Acetic acid, pentyl ester	16.68	0.66	0.69	0.69	0.71
Hexanoic acid, methyl ester	17.11	0.04	0.05	0.06	0.07
2-methyl-3-Heptanone	17.61	0.11	0.10	0.09	0.12
Trans-2-Heptenal	18.45	0.14	0.21	0.30	0.45
Benzaldehyde	18.82	0.15	0.17	0.16	0.14

Table 9. *(Continued)*

VOLATILE COMPOUND	RT (min)	7th DAY	21th DAY	35th DAY	49th DAY
2,6-dimethyl-4-Heptanone	18.91	0.11	0.09	0.12	0.13
6-methyl-5-hepten-2-one	19.42	1.22	1.25	1.23	1.22
Furan, 2-pentyl	19.67	0.17	0.15	0.14	0.18
Propanoic acid, pentyl ester	20.07	0.62	0.67	0.64	0.63
Hexyl acetate	20.42	0.07	0.07	0.08	0.09
Butanoic acid, 3-methyl, -butyl	21.38	0.51	0.53	0.52	0.55
1,2-dichloro-benzene	21.48	0.07	0.09	0.11	0.10
2-Octenal	22.08	0.32	0.35	0.33	0.33
N-Octanol	22.40	0.14	0.15	0.16	0.16
Butanoic acid, pentyl ester	23.14	0.22	0.22	0.21	0.18
Nonanal	23.62	0.24	0.23	0.20	0.21
Carvacrol	23.86	0.08	0.09	0.05	0.07
Cis-4-Decenal	26.39	0.07	0.04	0.05	0.06
Decanal	26.69	0.13	0.12	0.13	0.11
β-Cyclocitral	27.28	0.09	0.08	0.05	0.06
Geranial	28.17	0.09	0.08	0.08	0.09
Trans, trans 2,4-Decadienal	28.75	0.08	0.07	0.06	0.07
Eugenol	29.94	0.08	0.09	0.10	0.11
Geranyl acetone	31.42	0.34	0.38	0.43	0.57
β-ionone	32.08	0.04	0.05	0.04	0.06

Table 10. Volatile components of tomato (mg/kg) after exposure to 2 ppm of ozone gas packed in LDPE bags.

VOLATILE COMPOUND	RT (min)	7th DAY	21th DAY	35th DAY	49th DAY
Ethanol	2.33	0.50	0.53	0.73	0.98
Butanal	2.83	0.08	0.09	0.07	0.07
Acetic acid, ethyl ester	4.73	0.12	0.15	0.14	0.14
1-penten-3-one	7.36	0.25	0.35	0.54	0.65
Pentanal	7.83	0.20	0.21	023	0.22
3-methyl-1-butanol	9.40	0.18	0.18	0.19	0.20
Trans-2-pentenal	10.27	0.10	0.15	0.19	0.20

Table 10. *(Continued)*

VOLATILE COMPOUND	RT (min)	7th DAY	21th DAY	35th DAY	49th DAY
IS. 1-pentanol	10.80	20.25	20.25	20.25	20.25
N-Hexanal	12.17	2.33	2.86	3.11	3.72
Cis-3-hexenol	14.54	1.29	0.74	0.46	0.33
1-Hexanol	14.94	0.65	0.66	0.65	0.67
Cis-4-Heptenal	16.22	0.05	0.06	0.06	0.07
N-Heptanal	16.31	0.06	0.07	0.05	0.06
Acetic acid, pentyl ester	16.68	0.69	0.65	0.68	0.67
Hexanoic acid, methyl ester	17.11	0.04	0.05	0.05	0.04
2-methyl-3-Heptanone	17.61	0.12	0.13	0.14	0.15
Trans-2-Heptenal	18.45	0.19	0.26	0.38	0.46
Benzaldehyde	18.82	0.17	0.15	0.14	0.15
2,6-dimethyl-4-Heptanone	18.91	0.13	0.15	0.17	0.16
6-methyl-5-hepten-2-one	19.42	1.23	1.27	1.28	1.24
Furan, 2-pentyl	19.67	0.18	0.17	0.16	0.15
Propanoic acid, pentyl ester	20.07	0.83	0.77	0.76	0.75
Hexyl acetate	20.42	0.08	0.09	0.07	0.07
Butanoic acid, 3-methyl, -butyl	21.38	0.55	0.54	0.53	0.52
1,2-dichloro-benzene	21.48	0.10	0.09	0.09	0.08
2-Octenal	22.08	0.36	0.33	0.31	0.34
N-Octanol	22.40	0.12	0.13	0.14	0.15
Butanoic acid, pentyl ester	23.14	0.23	0.25	0.27	0.21
Nonanal	23.62	0.17	0.19	0.20	0.18
Carvacrol	23.86	0.07	0.08	0.06	0.07
Cis-4-Decenal	26.39	0.10	0.09	0.08	0.09
Decanal	26.69	0.11	0.13	0.11	0.12
B-Cyclocitral	27.28	0.08	0.05	0.06	0.04
Geranial	28.17	0.13	0.10	0.09	0.10
Trans, trans 2,4-Decadienal	28.75	0.08	0.09	0.08	0.07
Eugenol	29.94	0.08	0.09	0.10	0.11
Geranyl acetone	31.42	0.36	0.39	0.44	0.56
β-ionone	32.08	0.06	0.05	0.04	0.07

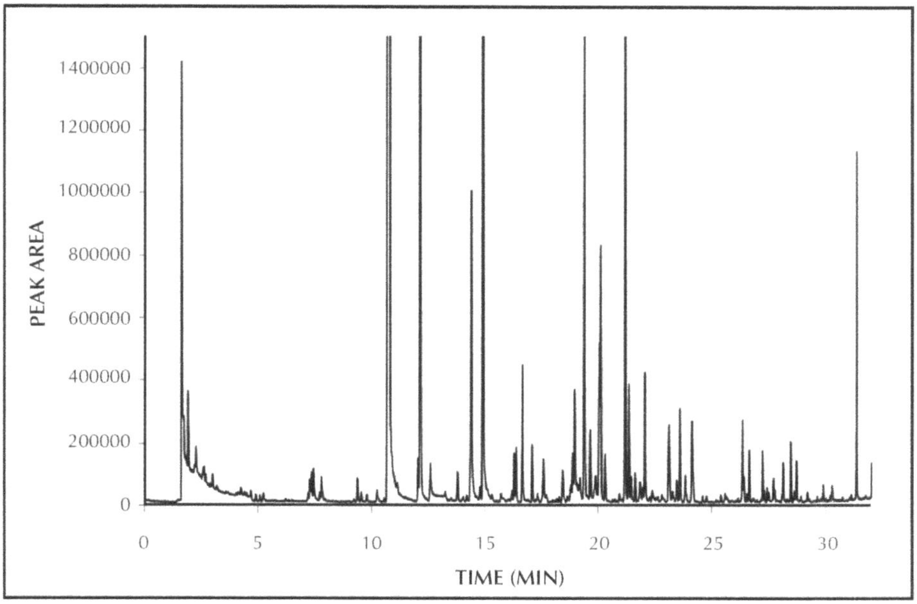

Figure 7. Chromatogram of tomato volatile components.

Sensory Evaluation

The effect of ozonation, packaging and storage time on tomato texture, taste, aroma, and overall appearance is shown in Tables 11, 12, 13, and 14 respectively. Sensory evaluation showed that the samples ozonated at concentrations of 0.5, 1, and 2 ppm packed either in commercial crates or in LDPE bags had better sensory characteristics compared to control samples. Texture, aroma, taste, and overall appearance scores of the fruit were retained above the lower acceptability limit in all ozonated samples throughout storage while control samples reached the lower limit of acceptability after 35 days.

Table 11. Effect of ozonation, packaging, and storage time on tomato texture.

DAYS	TEXTURE						
	control	0.5 ppm	0.5 ppm-LDPE	1 ppm	1 ppm-LDPE	2 ppm	2 ppm-LDPE
0	5.0±0.4	5.0±0.5	5.0±0.4	5.0±0.4	5.0±0.5	5.0±0.4	5.0±0.4
7	5.0±0.4	5.0±0.5	5.0±0.5	5.0±0.4	5.0±0.4	5.0±0.5	5.0±0.4
14	4.8±0.3	5.0±0.5	5.0±0.4	5.0±0.4	5.0±0.5	5.0±0.4	5.0±0.4
21	4.2±0.4	4.8±0.4	4.8±0.4	5.0±0.4	5.0±0.4	5.0±0.4	5.0±0.4
28	3.6±0.2	4.4±0.4	4.5±0.4	4.6±0.5	4.7±0.5	4.8±0.5	5.0±0.5
35	3.1±0.2	3.7±0.3	4.0±0.4	4.1±0.4	4.2±0.5	4.3±0.5	4.4±0.5
42	–	3.6±0.4	3.7±0.4	3.8±0.4	4.0±0.4	4.1±0.4	4.1±0.4
49	–	3.3±0.4	3.3±0.4	3.5±0.4	3.6±0.4	3.7±0.4	3.8±0.4

Ozonation produced statistically significant changes ($p < 0.05$) in texture (Table 11) after day 21 for control samples, after day 28 for ozonated samples with 0,5 ppm, 0,5 ppm-LDPE and 1 ppm, and after day 35 for samples ozonated with 1 ppm-LDPE, 2 ppm and 2 ppm-LDPE. Storage time significantly affected texture after day 21 for control samples, after 28 days for 0.5 ppm, 0.5 ppm-LDPE, and 1 ppm samples and after 35 days for 1 ppm-LDPE, 2 ppm and 2 ppm-LDPE samples.

Taste (Table 12) was significantly affected ($p < 0.05$) by storage time in control samples after day 14. For ozonated samples statistically significant changes ($p < 0.05$) were observed after day 21 for 0.5 ppm, after day 28 for samples 0.5 ppm-LDPE and 1 ppm and after the 35th day for samples 1 ppm-LDPE, 2 ppm, and 2 ppm-LDPE. Ozonation dose significantly affected ($p < 0.05$) all samples after day 14.

Table 12. Effect of ozonation, packaging and storage time on tomato taste.

DAYS	TASTE						
	Control	0,5 ppm	0,5 ppm-LDPE	1 ppm	1 ppm-LDPE	2 ppm	2 ppm-LDPE
0	5.0±0.5	5.0±0.4	5.0±0.5	5.0±0.4	5.0±0.5	5.0±0.4	5.0±0.5
7	5.0±0.4	5.0±0.4	5.0±0.4	5.0±0.4	5.0±0.4	5.0±0.5	5.0±0.4
14	4.2±0.4	4.6±0.4	4.8±0.4	5.0±0.4	5.0±0.4	5.0±0.5	5.0±0.4
21	3.6±0.3	4.4±0.4	4.5±0.4	4.8±0.4	4.8±0.4	4.8±0.4	4.8±0.5
28	3.3±0.3	4.1±0.4	4.3±0.4	4.6±0.4	4.6±0.5	4.7±0.4	4.7±0.4
35	2.8±0.2	3.6±0.4	3.8±0.4	3.9±0.4	4.1±0.4	4.2±0.4	4.4±0.5
42	–	3.4±0.3	3.6±0.3	3.7±0.3	3.8±0.3	3.8±0.3	3.9±0.3
49	–	3.2±0.3	3.3±0.3	3.5±0.3	3.5±0.3	3.6±0.3	3.7±0.3

Aroma (Table 13) was significantly affected ($p < 0.05$) by ozonation dose for all samples after day 14. Ozonated samples stored either in commercial crates or packaged in LDPE bags maintained higher levels of aroma (as compared to control samples) throughout storage. Regarding storage time, statistically significant changes ($p < 0.05$) in aroma were observed after day 14 for control samples, after day 28 in samples ozonated with 0.5 ppm, 0.5 ppm-LDPE, 1 ppm, and after the day 35 for 1 ppm-LDPE, 2 ppm, and 2 ppm-LDPE treated samples. Control samples reached the lower limit of acceptability on day 35, while all other samples remained above this limit throughout storage.

Similarly, appearance of tomatoes (Table 14) deteriorated in all samples during storage. Ozonated samples retained acceptable appearance throughout storage. Storage time significantly affected ($p < 0.05$) tomato appearance after day 21 for control samples and samples ozonated at 0.5 ppm, 0.5 ppm-LDPE and 1 ppm and after day 28 for samples ozonated at 1 ppm-LDPE, 2 ppm and 2 ppm-LDPE. Ozonation dose significantly affected ($p < 0.05$) tomato appearance after day 21 of storage.

Sensory evaluation showed that tomatoes treated with gaseous O_3 at concentrations of 0.5, 1, and 2 ppm either in commercial crates or packaged in LDPE bags, maintained their sensory characteristics as compared to control samples. This may be attributed to: (i) the antimicrobial activity of O_3 and (ii) the packaging materials used (LDPE), creating a passive modified atmosphere.

Table 13. Effect of ozonation, packaging and storage time on tomato aroma.

DAYS	AROMA						
	control	0,5 ppm	0,5 ppm-LDPE	1 ppm	1 ppm-LDPE	2 ppm	2 ppm-LDPE
0	5.0±0.4	5.0±0.4	5.0±0.5	5.0±0.4	5.0±0.5	5.0±0.4	5.0±0.5
7	5.0±0.4	5.0±0.5	5.0±0.5	5.0±0.4	5.0±0.4	5.0±0.5	5.0±0.4
14	4.4±0.5	4.8±0.4	5.0±0.4	5.0±0.4	5.0±0.4	5.0±0.4	5.0±0.4
21	3.9±0.3	4.7±0.4	4.8±0.4	4.8±0.4	4.8±0.4	4.8±0.5	4.8±0.4
28	3.3±0.2	4.5±0.4	4.5±0.4	4.6±0.3	4.6±0.3	4.7±0.4	4.7±0.4
35	3.1±0.2	3.8±0.4	3.9±0.4	4.0±0.4	4.1±0.4	4.2±0.4	4.3±0.3
42	–	3.7±0.4	3.7±0.4	3.9±0.3	3.9±0.4	4.0±0.3	4.0±0.3
49	–	3.4±0.3	3.5±0.3	3.6±0.4	3.7±0.4	3.7±0.3	3.8±0.4

Table 14. Effect of ozonation, packaging, and storage time on tomato appearance.

DAYS	OVERALL APPEARANCE						
	control	0,5 ppm	0,5 ppm-LDPE	1 ppm	1 ppm-LDPE	2 ppm	2 ppm-LDPE
0	5.0±0.5	5.0±0.5	5.0±0.4	5.0±0.4	5.0±0.4	5.0±0.4	5.0±0.4
7	5.0±0.4	5.0±0.5	5.0±0.5	5.0±0.5	5.0±0.4	5.0±0.4	5.0±0.4
14	5.0±0.4	5.0±0.4	5.0±0.4	5.0±0.4	5.0±0.4	5.0±0.5	5.0±0.4
21	4.3±0.4	4.6±0.5	4.6±0.5	4.7±0.4	4.7±0.4	4.8±0.4	4.9±0.5
28	3.4±0.2	4.1±0.4	4.1±0.4	4.2±0.3	4.3±0.3	4.4±0.4	4.5±0.4
35	2.8±0.2	3.7±0.3	3.7±0.3	3.8±0.4	3.9±0.4	4.0±0.4	4.0±0.3
42	–	3.5±0.2	3.5±0.2	3.6±0.2	3.7±0.3	3.8±0.2	3.8±0.3
49	–	3.4±0.2	3.4±0.2	3.5±0.2	3.6±0.2	3.7±0.2	3.8±0.3

Tzortakis et al. (2007) reported that sensory evaluation of tomatoes revealed substantial differences between O_3-treated tomatoes (0.05-1.0 μmol/mol for 12 days) and control samples, with up to 96% of judges expressing this preference. Of those expressing a preference, the majority (70% of judges) preferred fruit subjected to low-dose ozonation (0.15 μmol/mol). Non-parametric analysis of panel judgments revealed tomato fruit to be significantly sweeter ($p < 0.001$) following storage in an ozone-enriched atmosphere.

According to Aguayo et al. (2006) the effect of storage time on all sensorial properties of tomatoes was significant in control and ozonated samples (4 ± 0.5 μl/l) during the first 5 days of storage. Appearance, taste, aroma and overall quality of sliced tomatoes decreased during the first 5 days of storage but thereafter remained stable until day 12 for all samples. After day 12, appearance and overall quality declined until the end of storage, while taste and aroma were retained until the end of storage. Texture decreased gradually. At the end of storage, the O_3-treated tomatoes had a good appearance and overall quality while in control fruit these parameters received scores below the limit of acceptability. However, O_3-treated tomato slices showed a loss of aroma, scoring under the limit of acceptability after 15 days of storage. After 12 and 15 days at 5°C, O_3-treated tomatoes maintained a better appearance and overall quality than control samples. In agreement with these findings, total volatile aroma compounds were reduced by O_3 in strawberries (Pérez et al., 1999) and blueberries (Song et al., 2003). These authors reported that O_3 could have a nonreversible effect on the ability of fruits to produce volatile compounds. In contrast, an increase in aroma compounds was reported by Maguire and Solverg (1980) in whole tomatoes and by Ewell (1950) in strawberries, treated with 3.7 and 2–3 μl/l O_3, respectively.

CONCLUSION

Ozonation resulted in better quality retention of fresh tomatoes in relation to control samples during storage for 49 days. Even better quality resulted when ozone gas was combined with packaging in LDPE bags. Ozonation (specifically ozonation at a concentration of 2 ppm for 1 hr combined with packaging in LDPE bags and cold storage) is suggested for shelf life extension of fresh tomatoes by least 15 days compared to control samples.

KEYWORDS

- **Ozonation**
- **Packaging**
- **Shelf life**
- **Tomatoes**

Chapter 7

Combined Effect of Essential Oils and Nisin on Shelf Life Extension of Chicken Meat Stored at 4°C

E. Chouliara and M.G. Kontominas

INTRODUCTION

The combined effect of essential oils (EOs): oregano (0.1% v/w) or thyme (0.5% v/w) and the bacteriocin nisin (500 IU/g) on shelf life extension of fresh chicken meat stored at 4°C was investigated over a 20 day storage period. The following parameters were monitored: microbiological (total viable counts (TVC), *Pseudomonas* spp., *Brochothrix thermosphacta*, Enterobacteriaceae and lactic acid bacteria (LAB)), physico-chemical (pH, thiobarbituric acid test (TBA), color) and sensory (odor and taste) attributes. Microbial populations were reduced up to 2.8 log cfu/g for a given sampling day (day 6), with the more pronounced effect being achieved by the combination of nisin and thyme essential oil at concentration of 0.5% v/w. The TBA values for all treatments remained lower than 1 mg malondialdehyde (MDA)/kg of meat throughout the 20 day storage period. The pH values varied between 6.4 (day 0) and 5.9 (day 20). The L* parameter values decreased with time. Color parameters a* and b* were not affected by all treatments. Finally, sensory analysis showed that both EOs at the concentrations used, resulted in a desirable odor and taste of the product. Based primarily on sensory evaluation and secondarily on microbiological data, shelf life was 6 days for the control samples and samples treated with nisin, 9 days for samples containing 0.1% oregano EO or nisin plus 0.1% oregano EO and 12 days for samples containing 0.5% thyme EO or nisin plus 0.5 % thyme EO.

The shelf life extension of fresh chicken meat products has been a major concern of the poultry industry given product perishability (Chouliara and Kontominas, 2006). At the same time consumer demands for more "natural" and "minimally processed" food (Cleveland et al., 2001) has led to considerable research involving natural preservatives such as EOs and/or naturally produced antimicrobial agents (bacteriocins) for food preservation (Chouliara and Kontominas, 2006; Gill and Holley, 2003).

Bacteriocins are antimicrobial proteins or peptides produced by bacteria which are being widely used to ensure safety and retain quality in a wide range of food applications (Cleveland et al., 2001). Nisin belongs to bacteriocins produced by LAB which have been shown to be safe and can thus be used as effective natural food preservatives. Nisin is effective against gram-positive bacteria, including those that form heat-resistant endospores (Thomas et al., 2000). Gram-negative bacteria are resistant to nisin because of their nisin-impermeable outer membrane, but if gram negative cells are exposed to sublethal treatments that damage their outer membrane, nisin can gain access to the cytoplasmic membrane and cause microbial inhibition (Kalchayand et al., 1992; Stevens et al., 1991). Nisin finds applications in hurdle technology, which utilizes synergies of combined treatments to effectively preserve food (Fang and Lin, 1994; Nilsson et al., 2000; Szabo and Cahill, 1998). In the United States nisin has been assigned a Generally Recognized as Safe (GRAS) status since 1988 (US Food and Drug Administration, 1988).

Several studies have demonstrated antimicrobial activity of EOs against foodborne pathogens (Burt, 2004) as well as improvement of the sensory characteristics and extension of the shelf life of foods (Botsoglou et al., 2003; Chouliara and Kontominas, 2006; Goulas and Kontominas, 2007). Several of the studies generally agree that EOs are slightly more active against gram-positive than gram-negative bacteria (Burt, 2004) which could be due to the outer membrane of gram-negative bacteria that restricts diffusion of hydrophobic compounds through its lipopolysaccharide layer (Ratledge and Wilkinson, 1988, Vaara, 1992). However in several other studies no obvious difference in sensitivity between gram-positive and gram-negative bacteria was observed using a number of commercially available EOs (Burt, 2004).

Oregano and thyme are some of the most characteristic spices of the Mediterranean cuisine and are both well known for their antioxidative and antimicrobial activity (Botsoglou et al., 2003; Burt, 2004; Chouliara et al., 2006; Goulas and Kontominas, 2007). The two phenols, carvacrol and thymol are the major components of oregano and thyme essential oil found in different concentrations, mainly responsible for the antimicrobial activity of oregano and thyme EOs (Adam et al., 1998, Juliano et al., 2000). Eventhough, EOs are considered as safe (GRAS) (Lambert et al., 2001), their use is often limited by the strong odor/taste they impart to foodstuffs. For this reason the preservative effect of EOs may be achieved by using lower concentrations in combination with other preservation technologies such as low temperature (Chouliara et al., 2006, Skandamis and Nychas, 2001,), low dose irradiation (Chouliara et al., 2005, Farkas, 1990,), high hydrostatic pressure (Devlieghere et al., 2004) and MAP (Chouliara et al., 2006; Goulas and Kontominas, 2007; Marino et al., 1999).

Based on the above, the objective of the present work was to study the combined effect of EOs (oregano or thyme) and nisin for extending the shelf life of fresh breast chicken meat using microbiological, chemical and sensory analyses.

MATERIALS AND METHODS

Sample Preparation

Fresh chicken breast meat was provided by a local poultry processing plant (Pindos S.A., Ioannina, Greece) within 2 hr after slaughter in insulated polystyrene boxes on ice. Chicken samples weighing ca. 300 g in the form of chunks of approximate dimensions 2 x 2 x 2 cm, were placed in low density polyethylene/polyamide/low density polyethylene (LDPE/PA/LDPE) barrier pouches, 25 x 20 cm in dimensions, 75 μm in thickness having an oxygen permeability of 52.2 $cm^3\ m^{-2}\ d^{-1}\ atm^{-1}$ at 75% relative humidity (RH), 25°C and a water vapor permeability of 2.4 $g\ m^{-2}\ d^{-1}$ at 100% RH, 25°C, measured using the Oxtran 2/20 oxygen permeability and the Permatran W/31 water vapor permeability testers (MOCON, Inc, MN, USA) respectively.

A stock solution of nisin (Nisaplin, 50 x 10^6 IU/g, Aplin & Barrett Ltd., Dorset, England) was prepared by dissolving nisin in 0.02N HCl. The solution was sterilized by passing through a 0.2 μm pore filter and sterile distilled water was added to obtain a final stock solution of 25,000 IU/ml. The solution was stored at –20°C. Prior the experiments, the stock solution was brought to room temperature and diluted in water to the desirable concentration.

The following lots of samples were prepared: Lot 1 comprised the control samples (aerobic packaging). Nisin was added to the second lot (at a concentration of 500 IU/g of meat). Oregano oil 99% (Kokkinakis S.A., Athens, Greece) was pipetted to the third lot so as to obtain concentrations equal to 0.1% v/w. Thyme oil 99% (Kokkinakis S.A., Athens, Greece) was pipetted to lot 4 (concentration 0.5% v/w). Lot 5 contained nisin (500 IU/g) plus 0.1% oregano oil while lot 6 contained nisin plus 0.5% thyme oil. In preliminary experiments it was shown that the maximum concentration of oregano and thyme oil resulting in acceptable chicken odor and taste was 0.1 and 0.5% v/w, respectively. The contents of all pouches were gently massaged by hand so as to achieve a homogeneous distribution of both nisin and EOs in the substrate. Pouches were heat-sealed using a BOSS model N_{48} thermal sealer (BOSS, Bad Homburg, Germany) and kept at 4°C. Sampling was carried out on days: 0, 3, 6, 9, 12, 15, and 20 of storage.

Microbiological Analysis

Chicken samples (25 g) were transferred aseptically into individual stomacher bags (Seward Medical, UK), containing 225 ml of sterile Buffered Peptone Water (BPW) solution (0.1%) and homogenized in a stomacher (Lab Blender 400, Seward Medical, UK) for 60 s. For each sample, appropriate serial decimal dilutions were prepared in BPW solution (0.1%). The amount of 0.1 ml of these serial dilutions of chicken homogenates was spread on the surface of dry media. The TVC were determined using Plate Count Agar (PCA, Merck code 1.05463, Darmstadt, Germany), after incubation for 3 days at 30°C. Pseudomonads were determined on cetrimide fusidin cephaloridine agar (Oxoid code CM 559, supplemented with SR 103, Basingstoke, UK) after incubation at 25°C for 2 days (Mead and Adams, 1977). *Brochothrix thermosphacta* was determined on streptomycin sulphate-thallus acetate-cycloheximide (actidione) agar, prepared from basic ingredients in the laboratory after incubation at 25°C for 3 days. For members of the family Enterobacteriaceae, 1.0 ml sample was inoculated into 15 ml of molten (45°C) violet red bile glucose agar (Oxoid code CM 485). After setting, a 10 ml overlay of molten medium was added and incubation was carried out at 37°C for 24 hr. The large colonies with purple haloes were counted. The LAB was determined on de Man Rogosa Sharpe medium (Oxoid code CM 361). After setting, a 10 ml overlay of molten medium was added and incubation was carried out aerobically at 25°C for 5 days.

All plates were examined visually for typical colony types and morphological characteristics associated with each growth medium. In addition, the selectivity of each medium was checked routinely by gram staining and microscopic examination of smears prepared from randomly selected colonies from all of the media.

Physicochemical Analysis

The pH value was recorded using a Metrohm, model 691, pH meter. Chicken samples were thoroughly homogenized with 10 ml of distilled water and the homogenate used for pH determination. The TBA was determined according to the method proposed by Pearson (1991). The TBA content was expressed as mg of MDA/kg chicken. Color determination was carried out on the surface of breast chicken meat using a Hunter Lab model DP-9000 optical sensor colorimeter (Hunter Associates Laboratory, Reston VA, USA) as described by Du et al., 2002.

Sensory Evaluation

After each sampling all chicken samples (ca. 100g) were kept in the freezer (–20°C) for sensory evaluation. After the end of the experiment all the samples were cooked in a microwave oven at high power (700W) for 4 min including time of defrosting. A panel of 51 untrained judges, graduate students of the Department of Chemistry, was used for sensory analysis (acceptability test). Panelists were asked to evaluate taste and odor of cooked samples. Along with the test samples, the panelists were presented with a freshly thawed chicken sample, stored at –20°C throughout the experiment, this serving as the reference sample. Acceptability of odor and taste was estimated using an acceptability scale ranging from 5 to 1, with 5 corresponding to a most liked sample and 1 corresponding to a least liked sample. A score of 3.5 was taken as the lower limit of acceptability (Chouliara et al., 2007).

Statistical Analysis

Experiments were replicated twice on different occasions with different chicken samples. Analyses were run in triplicate for each replicate (n = 2 x 3). Microbiological data were transformed into logarithms of the number of colony forming units (cfu/g) and were subjected to analysis of variance (ANOVA). Means and standard deviations were calculated and when F-values were significant at the $P < 0.05$ level, mean differences were separated by the Least Significance Difference (LSD) procedure (Steel and Torrie, 1980).

DISCUSSION AND RESULTS

Microbiological Changes

The TVC values for all different chicken meat treatments are given in Table 1. The initial TVC (day 0) value for the fresh chicken meat was ca. 4.3 log cfu/g, indicative of good quality chicken meat (Dawson et al., 1995). The TVC reached a value of 7 log cfu/g, considered as the upper microbiological limit for fresh poultry meat, as defined by the ICMSF (1986), on day 6 for the air packaged samples and samples treated with nisin (500 IU/g); day 9 for samples containing 0.1% oregano oil and samples containing nisin plus 0.1% oregano EO and day 15 for both samples containing 0.5% thyme EO and nisin plus 0.5% thyme EO. A first observation to be made is that nisin did not affect TVC. On the contrary, oregano EO and its combination with nisin had a small but statistically insignificant ($P > 0.05$) effect on TVC while both thyme EO 0.5% and its combination with nisin had a clear effect in reducing TVC from 7.2 (air packaged samples) to 4.5 and 4.4 log cfu/g respectively on day 6 of storage ($P < 0.05$).

Table 1. Effect of EOs and nisin on TVC of chopped chicken meat stored at 4°C.

days of storage/treatment	0	3	6	9	12	15	20
	4.28*	5.83	7.19	7.55	7.78		
air packaged	(±0.20**)	(±0.31)	(±0.34)	(±0.31)	(±0.39)	ND	ND
	4.28	5.71	7.21	7.55	7.90		
air packaged + nisin	(±0.20)	(±0.24)	(±0.32)	(±0.22)	(±0.41)	ND	ND
	4.28	5.59	6.92	7.29	7.78	7.98	
air packaged +0.1% oregano oil	(±0.20)	(±0.23)	(±0.36)	(±0.31)	(±0.44)	(±0.45)	ND
	4.28	3.56	4.46	5.69	6.42	7.11	7.42
air packaged +0.5% thyme oil	(±0.20)	(±0.17)	(±0.26)	(±0.22)	(±0.31)	(±0.38)	(±0.44)
air packaged + nisin +0.1% oregano oil	4.28 (±0.20)	5.53 (±0.20)	6.95 (±0.33)	7.28 (±0.40)	7.81 (±0.44)	7.87 (±0.45)	ND
air packaged + nisin +0.5% thyme oil	4.28 (±0.20)	3.52 (±0.18)	4.37 (±0.24)	5.53 (±0.28)	6.38 (±0.35)	7.18 (±0.31)	7.35 (±0.43)

* log cfu/g ** S.D.
ND: not determined

The *Pseudomonads* (Table 2) are strictly aerobic gram-negative bacteria, one of the main spoilage microorganisms in fresh meat (Jay et al., 2005). Nisin had no effect ($P > 0.05$) on the reduction of the *Pseudomonads* population which can be explained by the presence of the outer membrane that gram negative bacteria possess which is impermeable to nisin. The EOs in turn, showed a clear concentration dependent antimicrobial activity against the *Pseudomonads*. This effect may be due to the ability of carvacrol, thymol, but also of p-cymene and γ-terpinene, minor constituents of both thyme and oregano oils, to penetrate the outer membrane of gram negative bacteria and cause microbial damage. On day 6 of storage oregano EO (0.1%) and thyme EO (0.5%) reduced the *Pseudomonads* population by 1.3 and 3.1 log cfu/g respectively ($P < 0.05$). The EO/nisin combinations did not show a synergistic effect behaving exactly like individual EOs. Thus the well documented (Kalchayanand et al., 1992, Stevens et al., 1991,) inhibition effect of the nisin plus EO combination on gram-negative bacteria was not shown in the case of the *Pseudomonads*.

Chouliara et al. (2007) reported a reduction in *Pseudomonads* higher than 5.4 log cfu/g in chicken meat with the addition of 1 % oregano oil after 6 days of storage. In contrast, Skandamis et al., (2002) reported that the *Pseudomonads* were the most resistant bacterial group to oregano oil. Likewise, Ouattara et al., (1997), reported low antimicrobial effects of oregano oil on a number of meat spoilage bacteria such as *Pseudomonas fluorescens*, *Brochothrix thermosphacta* and *Lactobacillus sakei*. Regarding the use of thyme EO, Deans and Richie (1987) showed that thyme oil, an essential oil containing similar components as oregano oil, was very effective against *Pseudomonas aeruginosa*, in study where the inhibitory properties of ten plant EOs were tested using an agar diffusion technique. Finally, Chouliara and Kontominaas

(2006) reported a reduction in *Pseudomonads* equal to 3.8 log cfu/g in chopped chicken meat with the addition of 1% thyme oil after 6 days of storage.

Table 2. Effect of essential oils and nisin on *Pseudomonas* spp. populations of chopped chicken meat stored at 4°C.

days of storage/treatment	0	3	6	9	12	15	20
air packaged	3.38* (±0.14**)	4.91 (±0.28)	6.28 (±0.24)	7.21 (±0.37)	7.46 (±0.42)	ND	ND
air packaged + nisin	3.38 (±0.14)	4.83 (±0.23)	6.73 (±0.24)	7.32 (±0.33)	7.49 (±0.39)	ND	ND
air packaged +0.1% oregano oil	3.38 (±0.14)	4.21 (±0.19)	4.92 (±0.26)	5.38 (±0.30)	6.49 (±0.39)	7.35 (±0.34)	ND
air packaged +0.5% thyme oil	3.38 (±0.14)	2.21 (±0.10)	3.18 (±0.17)	4.42 (±0.21)	5.71 (±0.32)	6.87 (±0.36)	7.31 (±0.40)
air packaged + nisin +0.1% oregano oil	3.38 (±0.14)	4.32 (±0.20)	4.75 (±0.21)	5.71 (±0.32)	6.20 (±0.35)	7.19 (±0.29)	ND
air packaged + nisin +0.5% thyme oil	3.38 (±0.14)	2.65 (±0.11)	3.09 (±0.17)	4.30 (±0.20)	5.63 (±0.24)	6.52 (±0.37)	7.10 (±0.32)

* log cfu/g ** S.D.
ND: not determined

Brochothrix thermosphacta is a gram-positive facultative anaerobic bacterium comprising part of the natural microflora of fresh meat packaged either aerobically or under MAP (Labadie, 1999). On day 6 of storage *B. thermosphacta* counts (Table 3) reached the value of 6.9 log cfu/g for aerobically packaged samples. On the same day *B. thermosphacta* counts were reduced by 1.8 log cfu/g by nisin 500 IU/g, 1.2 log cfu/g by oregano oil 0.1%, 4.2 log cfu/g by thyme oil (0.5%), 2.3 log cfu/g by nisin plus oregano EO (0.1%) and 4.6 log cfu/g by nisin plus thyme oil 1% ($P < 0.05$). Thus EOs and nisin showed an additive antimicrobial effect against *Brochothrix thermosphacta* with thyme oil exhibiting a higher antimicrobial effect than nisin most probably due to its higher concentration.

The above results are in agreement with those of Skandamis et al. (2002), who reported a reduction of *B. thermosphacta* by 1–2 log cfu/g in beef containing 0.8% oregano oil under different packaging conditions. Chouliara and Kontominas (2006) reported a reduction of 4.6 log cfu/g in *Brochothrix thermosphacta* counts with the addition of 1% thyme oil in chicken meat after 6 days of storage at 4°C. Cutter and Siragusa (1996) showed that spraying of vacuum packaged beef meat with nisin resulted to a population reduction of inoculated *Brochothrix thermosphacta* by 3 log cfu/g after a storage period of 21 days.

With respect to the Enterobacteriaceae, considered to be a hygiene indicator (Zeitoun et al., 1994), the initial counts were 2.3 log cfu/g (Table 4), indicative of good manufacturing practices in the poultry processing plant. As expected, nisin had no effect on the gram-negative Enterobacteriaceae. With the exception of day 3, oregano EO had also no effect on Enterobacteriaceae counts probably due to its low concentration. On the contrary thyme EO had a significant effect ($P < 0.05$) on Enterobacteriaceae populations. More specifically, thyme EO (0.5%) resulted to a reduction of 4.5 log cfu/g in chicken meat on day 6 of storage. The combination of thyme EO (0.5%) plus nisin showed an additive effect reducing the Enterobacteriaceae populations to lower than the method detection limit (1 log cfu/g) on day 6 of storage. Thus, the synergistic effect of nisin plus thyme combination was exhibited in the case of Enterobacteriaceae. The difference observed between the Pseudomonads and the Enterobacteriaceae requires further investigation.

Table 3. Effect of essential oils and nisin on *Brochothrix thermosphacta* populations of chopped chicken meat stored at 4°C.

days of storage/treatment	0	3	6	9	12	15	20
air packaged	3.04* (±0.13**)	5.17 (±0.22)	6.90 (±0.41)	7.23 (±0.31)	7.52 (±0.42)	ND	ND
air packaged + nisin	3.04 (±0.13)	3.60 (±0.15)	5.11 (±0.27)	5.30 (±0.24)	6.12 (±0.29)	6.96 (±0.36)	ND
air packaged +0.1% oregano oil	3.04 (±0.13)	4.39 (±0.20)	5.67 (±0.27)	6.48 (±0.29)	6.87 (±0.22)	7.41 (±0.39)	ND
air packaged +0.5% thyme oil	3.04 (±0.13)	3.00 (±0.17)	2.69 (±0.11)	4.18 (±0.18)	5.72 (±0.24)	6.18 (±0.33)	7.37 (±0.41)
air packaged + nisin +0.1% oregano oil	3.04 (±0.13)	3.51 (±0.15)	4.6 (±0.24)	4.92 (±0.22)	5.21 (±0.28)	5.58 (±0.23)	ND
air packaged + nisin +0.5% thyme oil	3.04 (±0.13)	<2.00	2.30 (±0.12)	2.82 (±0.19)	3.13 (±0.13)	3.87 (±0.16)	5.48 (±0.31)

* log cfu/g ** S.D.
ND: not determined

Present results are in general agreement with those of Chouliara and Kontominas (2006), who reported that thyme oil at a concentration of 1% had a strong effect in reducing Enterobacteriaceae counts from 6.2 to <1 log cfu/g in chicken meat after 6 days of storage at 4°C. In contrast, Skandamis and Nychas (2001) reported a reduction of Enterobacteriaceae counts by 2 log cfu/g in minced beef meat treated with 0.1% oregano oil under modified atmosphere packaging. Finally Chouliara et al. (2007) reported a reduction in Enterobacteriaceae population by at least 5.2 log cfu/g with the addition of 1% oregano oil in chicken meat after 6 days of storage.

The LAB are gram-positive bacteria behaving as facultative anaerobes, that can grow under different packaging conditions (in the presence or the absence of oxygen) and thus constitute a substantial part of the natural microflora of meats (Jay et al., 2005). The initial LAB counts (Table 5) were ca. 3.7 log cfu/g reaching 6.4 log cfu/g on day 6 of storage for the air packaged samples. Use of nisin resulted to a reduction of LAB by 1.9 log cfu/g on day 6 of storage ($P < 0.05$). On the same day, the use of oregano oil (0.1%) resulted to a reduction in LAB counts by 1.1 log cfu/g ($P < 0.05$), while thyme oil (0.5%) as well the combination of nisin plus thyme EO completely inhibited the growth of LAB. The combination of nisin and oregano EO (0.1%) resulted to a reduction of LAB counts by ca. 2.6 log cfu/g ($P < 0.05$). Thus, both EOs showed an additive effect with nisin against LAB. These findings, with regard to oregano oil, are in general agreement with the results of Zaika et al. (1983), who reported a LAB reduction of 4 log cfu/g in a pure culture after the addition of oregano oil at a concentration of 4 g/l. Such a concentration is ca. four times that used in the present study and also reflects differences in antimicrobial activity of EOs in model and food systems respectively (Burt, 2004). Present results with regard to thyme oil are in general agreement with those of Chouliara and Kontominas (2006) who reported a 5.4 log cfu/g reduction in LAB for chicken meat with the addition of 1% thyme oil after 6 days of storage. Giatrakou et al. (2010) also reported a 1 log cfu/g reduction in LAB for ready to eat chicken products with the addition of 0.2% thyme on day 6 of storage.

Table 4. Effect of essential oils and nisin on Enterobacteriaceae populations in chopped chicken meat stored at 4°C.

days of storage/treatment	0	3	6	9	12	15	20
air packaged	2.28* (±0.10**)	3.99 (±0.22)	6.15 (±0.34)	7.48 (±0.31)	7.59 (±0.42)	ND	ND
air packaged + nisin	2.28 (±0.10)	3.97 (±0.19)	6.25 (±0.36)	7.44 (±0.41)	7.59 (±0.32)	ND	ND
air packaged +0.1% oregano oil	2.28 (±0.10)	2.6 (±0.24)	5.81 (±0.32)	7.21 (±0.38)	7.51 (±0.35)	7.69 (±0.38)	ND
air packaged +0.5% thyme oil	2.28 (±0.10)	1.30 (±0.05)	1.61 (±0.09)	2.32 (±0.11)	3.45 (±0.20)	4.82 (±0.27)	6.03 (±0.32)
air packaged + nisin +0.1% oregano oil	2.28 (±0.10)	2.75 (±0.17)	5.90 (±0.32)	7.05 (±0.39)	7.65 (±0.43)	7.71 (±0.41)	ND
air packaged + nisin +0.5% thyme oil	2.28 (±0.10)	<1.00	<1.00	2.23 (±0.09)	3.51 (±0.20)	4.67 (±0.21)	5.85 (±0.32)

* log cfu/g ** S.D.
ND: not determined

The results of the present work showed that oregano oil at a concentration of 0.1% had a substantially lower effect than thyme oil (0.5%) in reducing the populations of bacterial groups studied. Such differences in antimicrobial activity between the two EOs, as previously stated, may be related to differences in concentration of oregano EO and thyme EO used in the present study.

Sensory Changes

Sensory properties (odor and taste) of cooked chopped chicken meat are given in Table 6. The lower acceptability limit of 3.5 was reached for odor after ca. 6 days, for both the control and nisin containing samples, 12 days for the samples containing 0.1% oregano EO and nisin plus 0.1% oregano EO and 15 days for samples containing thyme EO (0.5%) and nisin plus thyme EO. Taste proved to be a slightly more sensitive sensory attribute as shown by sensory scores in the Table 6. The lower acceptability limit for taste was reached after ca. 6 days for both the control and nisin containing samples, 9 days for samples containing 0.1% oregano EO and nisin plus oregano EO and 12 days for samples containing 0.5 % thyme EO and nisin plus thyme EO. Samples containing either 0.1% oregano oil or 0.5 % thyme oil gave a characteristic desirable odor and taste to the meat, very compatible to cooked chicken flavor. Present sensory data are in good agreement with those of Chouliara et al. (2007) and Chouliara and Kontominas (2006) regarding the effect of oregano and thyme oils on shelf life of chicken breast meat.

Sensory data were in reasonable agreement with microbiological data (TVC). Discrepancies between the two have been often observed and may be attributed to either the fact that it is not the total number of microorganisms but rather than the number of specific spoilage organisms (SSO) that cause spoilage (Jay et al., 2005) or the "masking" effect of EOs on deteriorative off-flavors produced during the spoilage of the food substrate.

Table 5. Effect of essential oils and nisin on LAB populations of chopped chicken meat stored at 4 °C.

days of storage/treatment	0	3	6	9	12	15	20
air packaged	3.66* (±0.21**)	5.04 (±0.27)	6.41 (±0.34)	7.02 (±0.39)	7.36 (±0.31)	ND	ND
air packaged + nisin	3.66 (±0.21)	3.69 (±0.14)	4.48 (±0.22)	5.29 (±0.29)	6.13 (±0.34)	ND	ND
air packaged +0.1% oregano oil	3.66 (±0.21)	4.93 (±0.26)	5.28 (±0.19)	6.17 (±0.35)	7.28 (±0.29)	7.36 (±0.31)	ND
air packaged +0.5% thyme oil	3.66 (±0.21)	<1.00	<1.00	2.81 (±0.11)	4.79 (±0.26)	6.21 (±0.33)	7.41 (±0.40)
air packaged + nisin +0.1% oregano oil	3.66 (±0.21)	2.95 (±0.13)	3.80 (±0.23)	5.08 (±0.28)	5.31 (±0.19)	5.84 (±0.30)	ND
air packaged + nisin +0.5% thyme oil	3.66 (±0.21)	<1.00	<1.00	1.19 (±0.06)	2.48 (±0.14)	4.61 (±0.27)	5.82 (±0.31)

* log cfu/g ** S.D.
ND: not determined

Physicochemical Changes

Changes in pH during the 20 day storage period for each different treatment, were small but statistically significant ($P < 0.05$) (data not shown). The pH values varied between 6.4 (day 0) and 5.9 (day 20) for aerobically packaged samples, 6.4 and 6.0 for nisin treated samples and between 6.4 and 6.1 for nisin treated samples containing either of the two EOs. Decrease in pH may be related to the growth of LAB producing lactic acid.

The TBA values for all different treatments are given in Table 7. The TBA values varied between 0.21 and 0.98 mg MDA/kg meat corresponding to very low degree of lipid oxidation.

This range of values is in agreement with that of Kim et al. (2005) for raw turkey and pork meat after seven days of storage (0.1-0.7 mg MDA/kg of meat). They are also in good agreement with those of Chouliara et al. (2007) who reported MDA values between 0.25 and 0.82 mg MDA/kg of chicken breast meat treated with 1% oregano oil stored at 4°C for a period of 25 days and those of Chouliara and Kontominas (2006) who reported TBA values between 0.15 and 0.64 mg MDA/kg of chicken breast treated with 1% thyme oil stored at 4°C for 25 days. Both oregano and thyme EO did not significantly affect degree of lipid oxidation under present experimental conditions despite the documented antioxidant properties of both EOs. In contrast, El-Alim et al. (1999) reported a 10% reduction in TBARS in raw pork patties after the addition of 10 g/kg of dried thyme plant.

Table 6. Effect of essential oils and nisin on sensory properties (odor and taste) of chopped chicken meat stored at 4°C.

days of storage/treatment	0	3	6	9	12	15	20	25
odour								
air packaged	5.0* (±0.2**)	5.0 (±0.2)	3.5 (±0.2)	2.3 (±0.2)	NT	NT	NT	NT
air packaged + nisin	5.0 (±0.3)	5.0 (±0.2)	3.6 (±0.2)	2.6 (±0.2)	NT	NT	NT	NT
air packaged + oregano 0.1%	5.0 (±0.3)	5.0 (±0.3)	4.7 (±0.4)	3.7 (±0.3)	3.3 (±0.2)	3.5 (±0.2)	3.0 (±0.2)	NT
air packaged + thyme 0.5%	5.0 (±0.3)	5.0 (±0.3)	5.0 (±0.3)	4.5 (±0.3)	4.0 (±0.2)	3.2 (±0.2)	NT	NT
air packaged + nisin +0.1% oregano oil	5.0 (±0.3)	5.0 (±0.3)	4.9 (±0.2)	4.0 (±0.2)	3.6 (±0.2)	3.3 (±0.2)	NT	NT
air packaged + nisin +0.5% thyme oil	5.0 (±0.3)	5.0 (±0.2)	5.0 (±0.2)	4.8 (±0.2)	4.3 (±0.2)	3.5 (±0.2)	3.0 (±0.1)	NT
taste								
air packaged	5.0 (±0.2)	5.0 (±0.2)	NT	NT	NT	NT	NT	NT
air packaged + nisin	5.0 (±0.2)	5.0 (±0.2)	NT	NT	NT	NT	NT	NT

Table 6. *(Continued)*

days of storage/treatment	0	3	6	9	12	15	20	25
air packaged + oregano 0.1%	5.0 (±0.2)	5.0 (±0.2)	4.7 (±0.3)	NT	NT	NT	NT	NT
air packaged + thyme 0.5%	5.0 (±0.2)	5.0 (±0.2)	4.7 (±0.3)	3.9 (±0.4)	3.4 (±0.3)	2.8 (±0.2)	NT	NT
air packaged + nisin +0.1% oregano oil	5.0 (±0.2)	5.0 (±0.2)	4.3 (±0.2)	NT	NT	NT	NT	NT
air packaged + nisin +0.5% thyme oil	5.0 (±0.2)	5.0 (±0.2)	4.5 (±0.2)	4.3 (±0.2)	3.8 (±0.1)	3.0 (±0.1)	NT	NT

* sensory scale: 5-1
** S.D.
NT: not tested

Table 7. Effect of essential oils and nisin on degree of lipid oxidation (mg MDA/kg) in chopped chicken meat stored at 4°C.

days of storage/treatment	0	3	6	9	12	15	20
air packaged	0.28 (±0.01*)	0.31 (±0.01)	0.58 (±0.03)	0.33 (±0.02)			
air packaged + nisin	0.28 (±0.01)	0.44 (±0.02)	0.43 (±0.02)	0.33 (±0.02)			
air packaged + oregano 0.1%	0.28 (±0.01)	0.35 (±0.02)	0.25 (±0.01)	0.28 (±0.02)	0.46 (±0.03)		
air packaged + thyme 0.5%	0.28 (±0.01)	0.21 (±0.01)	0.57 (±0.04)	0.50 (±0.05)	0.98 (±0.06)	0.71 (±0.09)	
air packaged + nisin +0.1% oregano oil	0.28 (±0.01)	0.30 (±0.01)	0.32 (±0.02)	0.30B (±0.02)	0.43 (±0.02)	0.42 (±0.02)	
air packaged + nisin +0.5% thyme oil	0.28 (±0.01)	0.25 (±0.01)	0.49 (±0.03)	0.39 (±0.01)	0.88 (±0.02)	0.51 (±0.06)	0.23 (±0.01)

* S.D.

Table 8. Effect of essential oils and nisin on color of chopped chicken meat stored at 4°C.

days of storage/ treatment	0			12			20		
	L*	a*	b*	L*	a*	b*	L*	a*	b*
air packaged	49.5 (±2.48**)	3.3 (±0.17)	11.4 (±0.57)	45.2 (±2.19)	6.1 (±0.25)	11.3 (±0.55)	NT	NT	NT
air packaged + nisin	49.5 (±2.48)	3.3 (±0.17)	11.4 (±0.57)	44.4 (±2.66)	3.6 (±0.18)	10.4 (±0.53)	NT	NT	NT

Table 8. *(Continued)*

air packaged + oregano 0.1%	49.5 (±2.48)	3.3 (±0.17)	11.4 (±0.57)	47.0 (±3.00)	4.5 (±0.38)	12.3 (±0.71)	NT	NT	NT
air packaged + thyme 0.5%	49.5 (±2.48)	3.3 (±0.17)	11.4 (±0.57)	44.4 (±4.51)	4.6 (±0.28)	14.2 (±1.31)	NT	NT	NT
air packaged + nisin +0.1% oregano oil	49.5 (±2.48)	3.3 (±0.17)	11.4 (±0.57)	44.7 (±2.24)	4.7 (±0.19)	11.9 (±0.18)	42.8 (±2.14)	3.0 (±0.15)	10.0 (±0.52)
air packaged + nisin +0.5% thyme oil	49.5 (±2.48)	3.3 (±0.17)	11.4 (±0.57)	50.7 (±2.54)	4.2 (±0.26)	14.1 (±0.71)	48.3 (±2.76)	4.1 (±0.21)	13.8 (±0.70)

** S.D.
NT: not tested

Likewise, Botsoglou et al. (2003) reported a threefold reduction in degree of lipid oxidation (0.6–0.2 mg MDA/kg) in raw turkey meat packaged aerobically after the addition of 200 mg/kg of oregano oil. Racanicci et al. (2004) reported a reduction in TBA values of 5 μmol MDA/kg of chicken meat balls after the addition of rosemary essential oil at a concentration of 0.05%. The reduction in TBA values further increased when the concentration of the essential oil was doubled. The TBA values less than or equal to 2 mg MDA/kg are considered to produce no adverse sensory changes in animal food substances (Byun et al., 2003).

As expected, nisin had no significant effect on TBA values when added alone to the chicken meat. The fact that the antioxidative effect of carvacrol and thymol, the two major constituents of oregano and thyme oil, was not shown in the present study may be due to the extremely low fat content of the samples (< 4%), greatly limiting the concentration of phenolic antioxidants in the fatty phase of the product, resulting inturn, to negligible differences in the degree of lipid oxidation of chicken meat as a function of storage time.

Color parameter values in raw chicken meat as affected by oregano or thyme oil and nisin on selected sampling days (day 0, 15, 25) are given in Table 8. Lightness (L* value) decreased in the samples packaged with nisin and EOs over the 20 day storage period. The product progressively became duller in color as also observed by Ahn et al. (1998). In turn, a* values, related to "redness" varied between 3.30 and 6.1 with no specific pattern produced by any of the treatments. Likewise, parameter b* related to "yellowness" of the product showed a variable trend with values ranging between 10.0 and 14.2 with no specific pattern produced by any of the treatments. Present color data are in general agreement with those of Chouliara et al. (2007) regarding the effect of oregano oil on L*, a* and b* values of chicken breast meat; they are also in agreement with those of Chouliara and Kontominas (2006) regarding the effect of thyme oil on color parameter of chicken breast meat.

CONCLUSION

Based on both sensory data (mainly taste) and microbiological data it can be concluded that nisin alone had no effect on shelf life of chopped chicken meat. The addition of 0.1% oregano and 0.5% thyme essential oil with or without nisin extended product shelf life by ca. 3 and 6 days respectively. Shelf life of control samples was ca. 6 days.

KEYWORDS

- **Chicken meat**
- **Nisin**
- **Oregano oil**
- **Shelf life extension**
- **Thyme oil**

Chapter 8

Combined Effect of Essential Oils and Nisin on the Survival of *Listeria monocytogenes* in Chicken Meat Packaged Aerobically at 4°C

E. Chouliara and M. G. Kontominas

INTRODUCTION

The combination of essential oils (EO) (oregano (OEO) or thyme (TEO)) 0.5% w/w and bacteriocin (nisin) at 500 IU/g was evaluated for the control of *Listeria monocytogenes* inoculated at 10^6 cfu/g onto raw chicken breast meat. Inoculated samples were aerobically packaged and stored at 4°C for a period of 10 days. Similar treatments of chicken breast samples were prepared without the addition of *L. monocytogenes* to determine sensory shelf life of products. Initial counts of *L. monocytogenes*, (6.23 log cfu/g) in chicken breast decreased by 4.15, 3.95, 3.09, >5.23 and >5.23 log cfu/g for samples treated with OEO 0.5%, TEO 0.5%, nisin 500 IU/g, OEO 0.5% plus nisin 500 IU/g, and TEO 0.5% plus nisin 500 IU/g, respectively on the second day of storage.

Treatment with the essential oils (oregano or thyme) was more effective in reducing *L. monocytogenes* population than the use of nisin at 500 IU/g. Sensory shelf life of chicken breast meat samples was ca. 6 days for the uninoculated samples, 8 days for samples treated with nisin and more than 10 days for samples with OEO, TEO, OEO plus nisin, and TEO plus nisin respectively. At the point of sensory rejection, the population of *L. monocytogenes* counts decreased by 2 (OEO 0.5%), 3.7 (TEO 0.5%), 3.4 (nisin), 4.2 (OEO plus nisin), and 3.7 (TEO plus nisin) log cfu/g.

Initial TVC (total viable count) (6.23 log cfu/g) of samples inoculated with *L. monocytogenes* decreased to 3.40 and 3.31 log cfu/g in the oregano oil plus nisin and the thyme oil plus nisin treated samples on day 2 of storage. After 10 days of storage TVC was under or equal to microbiological limit of 7 log cfu/g.

Initial pH (6.30) ranged between 6.67 (chicken plus *L. monocytogenes*) on day 10 of storage and 6.27 (chicken plus thyme oil plus nisin) on day 10 of storage implying changes in terms of putrefaction in the absence of essential oils and/or nisin.

According to the FAO/WHO (2002) report on foodborne diseases in Europe, approximately 26% of outbreaks caused by foods involved poultry and poultry products (including eggs). One of the greatest concerns of the poultry industry is the control of pathogens such as *Salmonella typhimurium*, *L. monocytogenes*, *E. coli* O157:H7, and *Staphylococcus aureus* contamination in the final product, attributed to cross contamination of poultry products during processing. Such contamination is highly probable due to the fact that poultry meat is an excellent substrate for the growth of both spoilage and pathogenic microorganisms. Of the above pathogens, *L. monocytogenes* can readily grow on poultry meat, surviving at low temperatures and reaches high populations under temperature abuse conditions. Such conditions may be encountered during the various stages of the distribution chain including warehouse storage, transportation, retail display, consumer handling and storage at home (Mytle et al., 2006).

Survival and growth of *Listeria* spp. may be controlled by the addition of natural preservatives such as essential oils and/or nisin.

Nisin is a polypeptide bacteriocin, produced by Lactic Acid Bacteria (LAB). It has been recognized as safe (GRAS) by the USFDA. It is the only bacteriocin approved for use as a natural food additive in over 50 countries (Jay, 2000). It is stable under refrigeration conditions, demonstrates heat stability, and is degraded in the digestive system. Nisin acts mainly against gram-positive bacteria by permeating the cytoplasmic membrane with the formation of transient pores, including those that form heat-resistant endospores (Thomas et al., 2000). Gram-negative bacteria are resistant to nisin because of their impermeable outer membrane, which prevents nisin to reach the cytoplasmic membrane. If, however, gram-negative cells are exposed to sublethal treatments which damage their outer membrane, nisin can gain access to the cytoplasmic membrane and cause microbial inhibition (Kalchayand et al., 1992, Stevens et al., 1991). Nisin finds applications in hurdle technology, which utilizes synergy of combined treatments to more effectively preserve food (Fang and Lin, 1994). Several studies have shown that nisin possesses antibacterial activity against *L. monocytogenes* in meat products (Montville and Chen, 1998, Vingolo et al., 2000).

Essential oils from spices, medicinal plants and herbs have been shown to possess antimicrobial properties and may serve as a source of natural antimicrobial agents against food pathogens (Deans and Ritchie, 1987, Kim et al., 1995) as well as for the extension of shelf life of foods (Botsoglou et al., 2003, Chouliara and Kontominas, 2006). More specifically, essential oils and their components are known to be active against wide variety of microorganisms, including gram-negative (Helander et al., 1998; Sivropoulou et al., 1996) and gram-positive bacteria (Kim et al., 1995). Oregano and thyme are some of the most characteristic spices of the Mediterranean cuisine, obtained by drying leaves and flowers of *Origanum vulgare* and *Thymus vulgaris* plants, respectively. They are both well known for their antioxidative and antimicrobial activity (Botsoglou et al., 2003, Burt, 2004). Carvacrol and thymol are the major components of oregano and thyme essential oil found in different concentrations and are mainly responsible for the antimicrobial activity of oregano and thyme essential oils (Adam et al., 1998, Juliano et al., 2000). Both essential oils have been found to be active against both gram-negative and particularly gram-positive bacteria (Shelef et al., 1980, Sivropoulou et al., 1996). Eventhough essential oils are considered as safe (GRAS) (Lambert et al., 2001), their use is often limited by the strong odor/taste they impart to foodstuffs. For this reason, the use of lower concentrations of essential oils may be combined with other preservation technologies such as low temperature (Chouliara and Kontominas, 2006), low dose irradiation (Chouliara et al., 2005, Farkas, 1990,), high hydrostatic pressure (Devlieghere et al., 2004) and MAP (Chouliara et al., 2007, Marino et al., 1999,).

The aim of the present work was to study the combined effect of essential oils (oregano or thyme) and nisin against *L. monocytogenes,* inoculated onto fresh chicken breast meat aerobically packaged under refrigerated storage.

MATERIALS AND METHODS

Preparation of Bacterial Inocula

A three-strain cocktail mixture of *Listeria* (two strains *L. monocytogenes* and one strain *L. inoccua*) was used to contaminate fresh chicken breast meat before packaging under aerobic conditions and storage. The strains were kept as frozen cultures in Tryptic Soy Broth (LAM M, Lancashire, UK) supplemented with 0.6 g Yeast Extract/100 ml (LAM M, Lancashire, UK) (TSBYE) plus 20% glycerol. The strains were activated by transferring 0.1 ml of the frozen culture into 10 ml of TSBYE and incubating at 30°C for 24 hr. Strains were subcultured twice (30°C for 24 hr) before use in the experiment. One ml of each strain was added in a sterile centrifuge tube and centrifuged for 15 min at 3,000 rpm, washed with phosphate buffered saline (PBS, LAM M, Lancashire, UK) and prior to addition to the samples, the cocktail mixture was diluted in Buffer Peptone Water (BPW, LAM M, Lancashire, UK) to a final concentration of 10^6 cfu/g of chicken meat.

Nisin Preparation

Nisin (Nisaplin 5×10^6 IU/g, Aplin & Barrett Ltd., Dorset, England) stock solution was prepared by dissolving nisin in 0.02 N HCl sterilized by filtering and added to sterile distilled water to a final concentration of 25,000 IU/ml. The stock solution was subsequently stored at -20°C. Before sample preparation, the stock solution was left at room temperature and diluted to a concentration of 500 IU.

Essential Oils

Both the essential oils of oregano and thyme (99% pure) were purchased from Kokkinakis S. A., Athens, Greece. Composition of both essential oils is shown in Table 1 (Karabagias et al., 2010).

Sample Preparation

Fresh chicken breast meat was provided by a local poultry processing plant (Pindos S.A., Ioannina, Greece) within 1 hr after slaughter in insulated polystyrene boxes on ice. Chicken samples, in chunks of approximate dimensions 2 x 2 x 2 cm, weighing ca. 300 g were placed in low density polyethylene/polyamide/low density polyethylene (LDPE/PA/LDPE) barrier pouches, 75 μm in thickness having an oxygen permeability of 52.2 $cm^3\ m^{-2}\ d^{-1}\ atm^{-1}$ at 75% relative humidity (RH), 25°C and a water vapor permeability of 2.4 $g\ m^{-2}\ d^{-1}$ at 100% RH, 25°C. Samples were inoculated with *Listeria monocytogenes* by pipeting the inoculum on the chicken meat surface so as to achieve a concentration of 10^5–10^6 cfu/g of chicken meat. Subsequently, samples were massaged by hand to uniformly distributed the inoculum on the meat surface. Following inoculation, chicken samples were kept at room temperature for 20 min to allow attachment. Subsequently samples were separated into 6 lots all inoculated with

L. monocytogenes. The first lot consisted of chicken meat inoculated with *Listeria* with no further treatment. Oregano and thyme oil were pipetted to the second and third lot of samples respectively so as to obtain a final concentration of 0.5% w/w. In the fourth lot nisin was added at a concentration of 500 IU/g. Lots 5 and 6 consisted of samples treated with both oregano oil (0.5% w/w) and nisin (500 IU/g) and both thyme oil (0.5% w/w) and nisin (500 IU/g), respectively. For comparison purposes control samples (chicken meat without addition of *L. monocytogenes*) were tested under identical experimental conditions.

Table 1. Composition of oregano and thyme oil (expressed as percentage of the total peak area of the chromatograms).

	Oregano oil	Thyme oil
Compounds	% composition	% composition
a-thujene	0.13	0.14
a-pinene	0.20	0.35
Camphene	0.07	0.14
1-octen-3-ol		0.10
b-myrcene	0.24	0.43
a-terpinene	0.43	0.68
p-cymene	14.97	11.91
limonene	0.12	0.14
b-phellandrene	0.08	0.12
1,8-cineole	0.09	0.12
g-terpinene	1.44	3.52
linalool	3.40	6.27
endo-borneol	0.20	0.41
4-terpineol	0.23	0.41
a-terpineol	0.15	0.24
thymol	21.89	54.55
carvacrol	55.76	12.29
caryophyllene	0.43	0.82
aromadendrene	0.09	0.16
caryophyllene oxide	0.09	0.20

Microbiological Analysis

Sampling of chicken samples was carried out at 0, 2, 4, 6, 8, and 10 days of storage. Twenty five g of chicken meat were transferred aseptically into individual stomacher bags (Seward Medical, UK), containing 225 ml of sterile Buffered Peptone Water (BPW) solution (0.1%) and homogenized in a stomacher (Lab Blender 400, Seward Medical, UK) for 60 s. For each sample, appropriate serial decimal dilutions were prepared in BPW solution (0.1%). The amount of 0.1 ml of these serial dilutions of chicken homogenates was spread in duplicate on TSAYE plates (non selective for enumeration of *L. monocytogenes* and other bacteria growing at 30°C) and PALCAM Agar (selective for enumeration of *L. monocytogenes*). For PALCAM agar 1 ml (0.25 ml × 4 times) of the first dilution of samples was also spread on four different plates in order to reduce the detection limit of the analysis to 1 log cfu/g instead of 2 log cfu/g. Colonies on agar plates were counted after incubation at 30°C for 2 days. For enrichment, 25 g sample was homogenized with 225 ml *Listeria* enrichment broth and incubated at 30°C for 24 and 48 hr for revival of injured cells. The enrichment broth was subcultured by surface plating on PALCAM agar incubated at 37°C for 24 and 48 hr and then examined for typical colonies. For members of the LAB 1.0 ml sample was inoculated into 15 ml of molten (45°C) de Man Rogosa Sharpe medium (Oxoid code CM 361) After setting, a 10 ml overlay of molten medium was added and incubation was carried out at 25°C for 3-4 days. *Pseudomonads* were determined on cetrimide fusidin cephaloridine agar (Oxoid code CM 559, supplemented with SR 103, Basingstoke, UK) after incubation at 25°C for 2 days. Microbiological counts were expressed as log cfu/g.

pH Measurement

The pH value was recorded using Metrohm, model 691, pH meter. Chicken samples were thoroughly homogenized with 10 ml of distilled water and the homogenate used for pH determination.

Sensory Analysis

The sensory characteristics of uninoculated chicken samples were evaluated. A panel composed of 51 untrained judges, faculty and graduate students of the Laboratory of Food Chemistry and Microbiology was used to evaluate chicken samples for taste and odor (acceptability study) using a 0-5 point scoring scale, with 5 being the most liked sample and 0 being the least liked sample (Chouliara et al., 2007). In a preliminary experiment it was shown that the highest acceptable concentration of both oregano and thyme oil was 0.5% (w/w). The lower acceptability limit was 3.5. The samples consisted of untreated chicken samples serving as controls, samples with oregano essential oil (0.5% w/w), samples with thyme oil (0.5% w/w), samples with nisin (500 IU/g), samples with nisin and oregano oil (0.5% w/w) and samples with nisin and thyme oil (0.5% w/w).

Statistical Analysis

Experiments were replicated twice on different occasions with different chicken samples. Analyses were run in triplicate for each replicate (n = 2 x 3). Microbiological data were transformed into logarithms of the number of colony forming units (cfu/g) and were subjected to analysis of variance (ANOVA). Means and standard deviations were calculated and when F-values were significant at the $P < 0.05$ level, mean differences were separated by the Least Significance Difference (LSD) procedure (Steel and Torrie, 1980).

DISCUSSION AND RESULTS

Sensory Analysis of Uninoculated Samples

Sensory scores (combined score for odor and taste) for uninoculated chicken samples are given in Table 2. The lower acceptability limit of 3.5 was reached for sensory properties (odor, taste) after 6 days, for the air packaged samples, 7 days for samples treated with nisin, and at least after 10 days for the samples containing 0.5% w/w OEO, 0.5% w/w TEO, 0.5% w/w OEO plus nisin, and 0.5% w/w TEO plus nisin. It should be noted that samples containing OEO plus nisin and TEO plus nisin received a relatively high score (3.7 and 4.3) even after 10 days of storage indicative of the strong additive effect of both essential oils plus nisin. Samples containing 0.5% w/w oregano and 0.5% w/w thyme oil gave a characteristic desirable odor and taste to the meat, very compatible to cooked chicken flavor.

Microbiological Changes

Enumeration of Listeria spp.

The combined effect of nisin, OEO or TEO on the fate of *Listeria monocytogenes* inoculated onto fresh chicken meat is shown in Table 3. The initial (day 0) population of *L. monocytogenes* was 6.2 log cfu/g while between day 6 and 10 it stabilized at 5.0 to 5.1 log cfu/g. The observed decrease is probably due to the antagonism of *Listeria* with other psychrotrophic bacteria such as *Pseudomonads* and LAB. Both essential oils (OEO and TEO) had substantial effect on *Listeria* spp. On the second day of storage counts of *Listeria* decreased to 2.1 and 2.3 log cfu/g, respectively, while the populations of *Listeria* in the control samples were 6.4 log cfu/g. The addition of nisin reduced the *Listeria* count to 3.1 log cfu/g (day 2 of storage). Thus both essential oils were more effective against *Listeria* compared to nisin. Interestingly in samples treated with both nisin and oregano or thyme oil the population of *Listeria* spp. was under the detection limit of 1 log cfu/g until day 4 of storage. It is worth mentioning that *Listeria* spp. was not detected on PALCAM plates in uninoculated chicken samples stored (data not showed). At the point of sensory rejection the *Listeria* count was: 5.0 log cfu/g for inoculated samples, 4.0 log cfu/g for nisin treated samples, 2.4 log cfu/g for OEO treated samples, 2.9 log cfu/g for TEO treated samples, 2.4 for OEO plus nisin treated samples and 2.7 for TEO plus nisin treated samples. Based on data in Table 1 it is most probable that the antimicrobial activity of OEO and TEO is due to carvacrol and thymol, the most abundant constituents in both essential oils. Secondary components of OEO and TEO such as para-cymene, γ-terpinene and linalool may also contribute to the antimicrobial effect of these essential oils.

Table 2. Combined effect of nisin, oregano and thyme essential oil on the sensory properties (combined score of odor and taste) of chopped chicken meat stored aerobically at 4°C.

Sample	0	2	4	6	8	10
	5.0	5.0	4.4	3.5	2.7	2.2
chicken	(±0.3*)	(±0.4)	(±0.3)	(±0.3)	(±0.2)	(±0.1)
	5.0	5.0	4.8	4.7	3.9	3.6
chicken + oregano oil 0.5%	(±0.3)	(±0.3)	(±0.4)	(±0.3)	(±0.3)	(±0.2)

Table 2. *(Continued)*

Sample	0	2	4	6	8	10
chicken + thyme oil 0.5%	5.0 (±0.3)	5.0 (±0.3)	5.0 (±0.2)	4.9 (±0.2)	4.5 (±0.3)	3.9 (±0.3)
chicken + nisin	5.0 (±0.2)	5.0 (±0.3)	4.6 (±0.3)	4.0 (±0.2)	3.2 (±0.1)	2.1 (±0.1)
chicken + oregano oil 0.5% + nisin	5.0 (±0.3)	5.0 (±0.2)	5.0 (±0.2)	4.9 (±0.2)	4.3 (±0.2)	3.7 (±0.1)
chicken + thyme oil 0.5% + nisin	5.0 (±0.3)	5.0 (±0.3)	4.9 (±0.3)	4.8 (±0.2)	4.6 (±0.2)	4.3 (±0.1)

Scoring scale: 0-5
* S.D.

The results of the above study are in general agreement with those of Solomakos et al. (2008), who reported a reduction of *L. monocytogenes* inoculated in minced meat by 0.4 (nisin 500 IU/g), 1.1 (nisin 1000 IU/g), 1.9 (thyme 0.6%), 2.5 (thyme and nisin 500 IU/g) and >3 log cfu/g (thyme and nisin 1000 IU/g). The results of the present study are also in general agreement with those of Ferreira and Lund (1996), who reported a decrease of 3 log cfu/g in *L. monocytogenes* population in cottage cheese spiked with 10^4 cfu/g *L. monocytogenes*, with the addition of 2000 IU/g nisin, after 7 days of storage at 20°C. According to Davies et al. (1997) the growth of *L. monocytogenes* inoculated at 10^2-10^3 cfu/ml in ricotta cheese was inhibited up to 55 days at 6-8°C when 2.5 mg/l nisin was added. Pawar et al. (2000) also reported that the degree of inhibition of *L. monocytogenes* in raw minced buffalo meat increased with increasing concentration of nisin from 400 to 800 IU/g during storage at 4°C.

Regarding the control of *L. monocytogenes*, numerous researchers believe that nisin is not effective in meat substrates due to product's high pH (Burt, 2004) and interference by meat components (de Vuyst and Vandamme, 1994) while others find contradictory results (Chung et al., 1989).

Other studies also reported oregano and thyme oils to be effective at levels of 5–20 µl/g in inhibiting *L. monocytogenes* and autochthonous spoilage flora in meat products (Aureli et al., 1992, Skandamis and Nychas, 2001, Stecchini et al., 1993, Tsigarida et al., 2000;).

Table 3. Combined effect of nisin, oregano or thyme essential oil on the counts of *Listeria* spp. inoculated at 10^6 cfu/g in chopped chicken meat, packaged aerobically at 4°C.

Sample	0	2	4	6	8	10
chicken inoculated with *Listeria* cocktail	6.23* (±0.37**)	6.36 (±0.25)	5.36 (±0.29)	5.02 (±0.28)	5.08 (±0.30)	5.06 (±0.29)
chicken inoculated with *Listeria* cocktail + oregano oil 0.5%	6.23 (±0.37)	2.08 (±0.12)	2.20 (±0.13)	2.20 (±0.12)	2.48 (±0.21)	2.38 (±0.19)
chicken inoculated with *Listeria* cocktail+ thyme oil 0.5%	6.23 (±0.37)	2.28 (±0.17)	2.51 (±0.18)	2.63 (±0.21)	2.85 (±0.21)	2.88 (±0.20)

Table 3. *(Continued)*

Sample	0	2	4	6	8	10
chicken inoculated with *Listeria* cocktail+ nisin	6.23 (±0.37)	3.14 (±0.16)	3.51 (±0.21)	3.96 (±0.23)	3.98 (±0.22)	4.22 (±0.24)
chicken inoculated with *Listeria* cocktail+ oregano oil 0.5% + nisin	6.23 (±0.37)	<1.00	<1.00	1.45 (±0.07)	2.08 (±0.17)	2.43 (±0.14)
chicken inoculated with *Listeria* cocktail+ thyme oil 0.5% + nisin	6.23 (±0.37)	<1.00	<1.00	1.80 (±0.08)	2.52 (±0.17)	2.66 (±0.15)

* log cfu/g
** S.D.

Enumeration of TVC

The TVC values for all different chicken meat treatments are given in Table 4. The initial value of TVC (day 0) for the fresh uninoculated chicken meat was ca. 4.3 log cfu/g and 6.3 log cfu/g for samples inoculated with *Listeria* spp. On day 6 of storage TVC reached a value of 7.2 and 7.0 log cfu/g for samples inoculated with *Listeria* spp. and uninoculated samples respectively. This is indicative of antagonistic growth between *Listeria* spp. and natural spoilage microflora of chicken. On the same day, TVC values were 5.6, 6.3, 7.0, 4.5, and 4.7 log cfu/g for treatments with OEO, TEO, nisin, OEO plus nisin. and TEO plus nisin, respectively. The combination of nisin and oregano or thyme oil had the most pronounced effect on reduction of TVC population.

The above results, regarding the use of nisin, are in general agreement with those of Yuste et al., 1998, who reported a reduction of 0.5 log cfu/g in the TVC of mechanically deboned chicken meat when treated with nisin at a concentration of 200 ppm. With respect to the use of oregano oil, present results are in general agreement with those of Tsigarida et al., (2000), who reported a reduction in initial microflora of beef meat fillets by 2-3 log cfu/g with the addition of 0.8% of oregano essential oil. They are also in agreement with those of Skandamis and Nychas (2001) who reported an immediate suppression of TVC in minced beef meat by 1 log cfu/g when oregano oil was added at a concentration of 1% v/w.

Table 4. Combined effect of nisin, oregano and thyme essential oil on TVC of chopped chicken meat inoculated with *Listeria* spp. (10^6 cfu/g), stored aerobically at 4°C.

Sample	0	2	4	6	8	10
uninoculated chicken	4.28* (±0.21**)	5.45 (±0.29)	6.26 (±0.36)	7.19 (±0.38)	7.31 (±0.32)	7.63 (±0.39)
chicken inoculated with *L. monocytogenes*	6.29 (±0.37)	6.78 (±0.39)	7.04 (±0.41)	6.98 (±0.38)	7.49 (±0.42)	7.93 (±0.43)
chicken inoculated with *L. monocytogenes*+ oregano oil 0.5%	6.29 (±0.37)	4.48 (±0.27)	5.02 (±0.29)	5.60 (±0.31)	5.93 (±0.36)	6.28 (±0.37)
chicken inoculated with *L. monocytogenes*+ thyme oil 0.5%	6.29 (±0.37)	5.35 (±0.31)	5.20 (±0.33)	6.25 (±0.31)	6.30 (±0.27)	6.77 (±0.32)
chicken inoculated with *L. monocytogenes*+ nisin	6.29 (±0.37)	5.42 (±0.32)	6.30 (±0.36)	6.95 (±0.38)	7.17 (±0.39)	7.23 (±0.34)

Table 4. *(Continued)*

Sample	0	2	4	6	8	10
chicken inoculated with *L. monocytogenes*+ oregano oil 0.5% + nisin	6.29 (±0.37)	3.20 (±0.19)	3.74 (±0.21)	4.50 (±0.21)	4.75 (±0.23)	5.04 (±0.32)
chicken inoculated with *L. monocytogenes*+ thyme oil 0.5% + nisin	6.29 (±0.37)	3.31 (±0.18)	3.92 (±0.19)	4.73 (±0.29)	4.94 (±0.30)	5.19 (±0.31)

* log cfu/g

** S.D.

Enumeration of Lactic Acid Bacteria

The LAB is fermentative bacteria that can grow in the presence or absence of oxygen, constituting a substantial part of the natural microflora of meats (Jay, 1986). The initial LAB counts for inoculated with *Listeria* spp. samples (Table 5) were ca. 3.7 log cfu/g (on day 0) and reached 7.5 log cfu/g on day 10 of storage for the air packaged samples. On the same day of storage (day 10) LAB populations were 5.3, 5.7, 5.7, 4.8, and 5.1 for samples treated with OEO, TEO, nisin, OEO plus nisin, and TEO plus nisin, respectively. It is obvious that essential oils showed an additive effect with nisin for the reduction of LAB populations. Based on the above data one may conclude that the lethal effect of both essential oils was mediocre as LAB are known to be quite resistant to such agents (Burt, 2004)

Zaika et al. (1983) reported a reduction of 4 log cfu/g in a pure culture of LAB after the addition of oregano oil at a concentration of 4 g/l.

Table 5. Combined effect of nisin, oregano and thyme essential oil on LAB of chopped chicken meat inoculated with *Listeria* spp. (10^6 cfu/g), stored aerobically at 4°C.

Sample	0	2	4	6	8	10
uninoculated chicken	3.66* (±0.31**)	4.82 (±0.31)	5.93 (±0.38)	6.41 (±0.44)	7.11 (±0.43)	7.52 (±0.39)
chicken inoculated with *L. monocytogenes*	3.68 (±0.31)	4.95 (±0.24)	7.08 (±0.36)	6.70 (±0.31)	7.05 (±0.37)	7.32 (±0.35)
chicken inoculated with *L. monocytogenes*+ oregano oil 0.5%	3.68 (±0.31)	4.18 (±0.22)	4.77 (±0.24)	5.02 (±0.25)	5.34 (±0.26)	5.32 (±0.26)
chicken inoculated with *L. monocytogenes*+ thyme oil 0.5%	3.68 (±0.31)	4.28 (±0.23)	4.98 (±0.21)	5.12 (±0.27)	5.41 (±0.27)	5.69 (±0.27)
chicken inoculated with *L. monocytogenes*+ nisin	3.68 (±0.31)	3.82 (±0.22)	4.11 (±0.19)	4.91 (±0.28)	5.11 (±0.24)	5.68 (±0.28)
chicken inoculated with *L. monocytogenes*+ oregano oil 0.5% + nisin	3.68 (±0.31)	1.48 (±0.05)	2.38 (±0.12)	3.22 (±0.16)	3.48 (±0.19)	4.77 (±0.24)
chicken inoculated with *L. monocytogenes*+ thyme oil 0.5% + nisin	3.68 (±0.31)	1.58 (±0.08)	2.64 (±0.11)	3.41 (±0.13)	4.52 (±0.21)	5.07 (±0.25)

* log cfu/g

** S.D.

Enumeration of *Pseudomonads'* Population

Pseudomonads are gram-negative, strictly aerobic bacteria comprising the main spoilage microorganism in meat products (Jay, 1986). The initial *Pseudomonads* counts (Table 6) for uninoculated samples were 3.4 log cfu/g (day 0) and reached 7.5 log cfu/g on day 10 of storage. On the same day (day 10) *Pseudomonads*' populations were 7.1, 5.3, 5.7, 5.7, 4.8, and 5.1 for samples inoculated with *Listeria* spp., samples treated with OEO, TEO, nisin, OEO plus nisin, and TEO plus nisin, respectively. As expected nisin did not affect *Pseudomonads*' growth while the use of both essential oils showed an obvious effect on the growth of these bacteria.

The findings of the above study are in general agreement with those of Deans and Richie (1987), who showed that thyme oil was the most effective against *Pseudomonas aeruginosa* in study where the inhibitory properties of ten plant essential oils were tested using an agar diffusion technique. In contrast, Skandamis et al. (2002) reported that the *Pseudomonads* were the most resistant group to oregano oil as compared to other spoilage flora.

pH Determination

Changes in pH (data not shown) during the 10 day storage period for each different treatment, were small but statistically significant ($p < 0.05$) only in the case of samples inoculated with *Listeria* spp. The pH values varied between 6.3 (day 0) for all samples and 6.7 (day 10) for samples inoculated with *Listeria* spp. Such changes in pH imply small changes in the substrate in terms of acid production (action of LAB) or alkaline products production (putrefaction).

Table 6. Combined effect of nisin, oregano and thyme essential oil on *Preudomonads'* population of chopped chicken meat inoculated with *Listeria* spp. (10^6 cfu/g), stored aerobically at 4°C.

sample	0	2	4	6	8	10
uninoculated chicken	3.38* (±0.18**)	4.83 (±0.34)	5.19 (±0.41)	6.28 (±0.50)	7.09 (±0.49)	7.47 (±0.45)
chicken inoculated with *L. monocytogenes*	3.35 (±0.31)	4.56 (±0.30)	4.92 (±0.29)	5.88 (±0.28)	6.92 (±0.31)	7.11 (±0.36)
chicken inoculated with *L. monocytogenes*+ oregano oil 0.5%	3.35 (±0.31)	<2.00	2.38 (±0.27)	3.46 (±0.23)	4.88 (±0.24)	5.26 (±0.28)
chicken inoculated with *L. monocytogenes*+ thyme oil 0.5%	3.35 (±0.31)	2.21 (±0.23)	4.98 (±0.25)	5.12 (±0.26)	5.41 (±0.25)	5.69 (±0.28)
chicken inoculated with *L. monocytogenes*+ nisin	3.35 (±0.31)	4.82 (±0.21)	5.11 (±0.19)	6.31 (±0.28)	7.11 (±0.24)	7.43 (±0.29)
chicken inoculated with *L. monocytogenes*+ oregano oil 0.5% + nisin	3.35 (±0.31)	<2.00	2.38 (±0.12)	3.22 (±0.16)	3.48 (±0.19)	4.77 (±0.24)
chicken inoculated with *L. monocytogenes*+ thyme oil 0.5% + nisin	3.35 (±0.31)	2.23 (±0.07)	2.64 (±0.13)	3.41 (±0.11)	4.52 (±0.19)	5.07 (±0.23)

* log cfu/g

** S.D.

CONCLUSION

The results of this study showed that inoculation of chicken meat with *Listeria* spp. which constitutes a serious public health problem may be partly controlled by the combined use of natural preservatives such as essential oils of oregano or thyme oil and nisin. The combination of nisin with either oregano or thyme oil resulted in the reduction of *Listeria* spp. under the detection limit (1 log cfu/g) in chicken samples for 4 days of storage at 4°C (<2 log cfu/g for 6 days of refrigerated storage). It would be interesting to study the effect of specific as well as other essential oils on the fate of *Listeria* spp. when inoculated into the food at a lower concentration (i.e. 10^3 cfu/g) corresponding to a more realistic level of contamination.

Regarding essential oils, it should be stressed that they may have a great potential for use as natural food preservatives. Spices and herbs, which are used as ingredients in prepared foods or added as flavoring agents to foods, can be also used for their antimicrobial properties. On the other hand, volatile oils must be added in small amounts to foods in order to prevent deterioration of foodstuffs' sensory properties. Additional research should be undertaken to confirm and quantify the antimicrobial effect of specific essential oils against individual pathogens in particular food substrates while at the same time to minimize the effect of essential oils on sensory properties of foodstuffs.

KEYWORDS

- **Chicken meat**
- ***Listeria monocytogenes***
- **Nisin**
- **Oregano essential oil**
- **Thyme essential oil**

Chapter 9

Shelf Life Extension of Ground Meat, Stored at 4°C using Chitosan and an Oxygen Absorber

N. Chounou, E. Chouliara, S. F. Mexis, D. Georgantelis, and M. G. Kontominas

INTRODUCTION

Shelf life extension of fresh ground meat stored at 4°C was investigated using either chitosan (1% w/w) or an oxygen absorber or the combination of the two. Aerobically packaged meat samples stored at 4°C were taken as controls. Parameters monitored over 10 day storage period were: microbiological (TVC, *Pseudomonas* spp., Lactic Acid Bacteria (LAB) and Enterobacteriaceae), physico-chemical (pH, TBA) and sensory (odor and taste) attributes. Microbial populations were reduced by 0.4-2.0 log cfu/g for a given sampling day, with the more pronounced effect being achieved by the combination of chitosan and the oxygen absorber. Of the two, the oxygen absorber proved more effective in reducing microbial populations. The TBA values for control samples, samples treated with chitosan and samples containing the oxygen absorber increased (0.88-2.9, 0.88-1.59 and 0.88-0.93 mg MDA/kg of meat respectively) during storage while for samples treated with both chitosan and the oxygen absorber TBA values decreased from 0.88 to 0.36 mg MDA/kg of meat. Changes in pH values for the different treatments were statistically insignificant ($p > 0.05$). Based on both sensory evaluation and microbiological analysis, shelf life of ground meat samples was: 3 days for control samples, 3-4 days for samples treated with 1% chitosan, 6 days for samples packaged with the oxygen absorber and 9 days for samples treated with 1% chitosan and packaged with the oxygen absorber.

Given the increasing consumer demand for minimally processed foods without chemical preservatives, the food industry is facing a constant challenge to develop alternative "natural" methods to extend product shelf life and improve safety (Helander, 2001). Meat and especially ground meat products are highly susceptible to both microbial growth and lipid oxidation due to their large surface to weight ratio, leading to rapid spoilage and development of rancid or warmed-over flavor respectively.

Active packaging refers to the incorporation of specific additives into the packaging film or container with the aim of maintaining quality and extending product shelf life (Day, 1989). The most widely used active packaging concepts are those developed to scavenge oxygen and was first commercialized in 1970 by Mitsubishi Gas Chemical Company (Japan). The purpose of the oxygen scavenger is to create infinitely low O_2 atmosphere within the pack, preventing deterioration through oxidation and growth of aerobic microorganisms (Mohan et al., 2008). Ageless is the most common O_2 absorber system based on iron (Fe^{2+}) oxidation (Nakamura and Hoshino, 1983). The Ageless sachets are designed to reduce O_2 levels to less than 0.01% (Labuza, 1987) in the package microenvironment. Oxygen absorbers have been used effectively to prevent discoloration of cured meats, rancidity problems in high-fat foods, mold growth in high moisture bakery products and so on. (Berenzon and Saguy, 1998). Besides the advantages, the use of O_2 absorbers has also certain disadvantages. An anoxic environment in the case of foods with water activity greater than 0.92 may enhance the growth of anaerobic pathogens including *Clostridium botulinum* and thus may introduce health risks if the temperature is not kept below or equal to 3°C (Mohan et al., 2008).

Chitosan is versatile biopolymer with a wide range of applications in the food industry (Rudrapatnam and Farooqahmed, 2003; Shahidi et al., 1999). It has been classified as generally recognized as safe by the US Food and Drug Administration (USFDA, 2001) and has broad-spectrum antimicrobial activity against both gram-positive and negative bacteria as well as fungi (Ouattara et al., 1997). The main mechanism of antimicrobial activity of chitosan is believed to be the binding of trace metals on the bacterial membrane affecting its permeability (Helander, 2001; Zheng and Zhu, 2003).

Recent studies have demonstrated the potential application of chitosan as a preservative in various meat products (Darmadji and Izumimoto, 1994; Lin and Chao, 2001; Ouattara et al., 1997; Sagoo et al., 2002; Soultos et al., 2008).

Chitosan can also be used to inhibit rancidity of muscle foods during storage. The mechanism by which this inhibition takes place is probably related to chelation of free iron that is released from meat hemoproteins during heat processing and storage (Georgantelis et al., 2007). The effect of chitosan on oxidative stability of minced beef was studied by Darmadji and Izumimoto (1994) who observed that the addition of chitosan (1%) resulted in 70% reduction of TBA values of meat after 3 days at 4°C.

Based on the above, the objective of the present work was to study the combined effect of chitosan plus an oxygen absorber on shelf life extension of ground meat stored at 4°C.

MATERIALS AND METHODS

Sample Preparation

Food grade chitosan (MW 120 kDa, degree of deacetylation 85%) was provided by Oligopharm Co. Ltd., Nizhni Novgorod, Russia. Fresh beef meat was provided by a local meat processing plant (SVEKI S. A., Ioannina, Greece) 24 hr post slaughter in chunks of dimensions 3 x 3 x 3 cm and transferred within 30 min to the laboratory in insulated polystyrene boxes on ice. Meat was ground in home type meat grinder presterilized using boiling water.

Four lots of samples were prepared: The first lot comprised the controls (aerobically packaged). Chitosan was added in the form of powder to the ground meat to the second lot so as to obtain a final concentration equal to 1% (w/w). The contents of the pouch were gently massaged by hand for the homogenous distribution of chitosan. Lot three consisted of samples in which a ZTP type Ageless® O_2 absorber sachet (Mitsubishi Gas Chemical Company, Japan) was added inside the package. Finally, the fourth lot consisted of samples in which both chitosan (1% w/w) and the ZTP type O_2 absorber were added to the meat and package respectively. Meat samples were placed in low density polyethylene/polyamide/low density polyethylene (LDPE/PA/LDPE) barrier pouches, 75 μm in thickness, having oxygen permeability of 52.2 $cm^3\ m^{-2}\ d^{-1}\ atm^{-1}$ at 75% relative humidity (RH), 25°C and water vapor permeability of 2.4 $g\ m^{-2}\ d^{-1}$ at 100% RH, 25°C. Pouches were heat sealed using BOSS model N48 thermal sealer (BOSS, Bad Homburg, Germany) and stored at 4°C.

The sampling was carried out on day 0, 2, 4, and 6 of storage for controls and chitosan treated samples and on day 0, 2, 4, 6, 8, and 10 of storage for the rest of the samples.

Microbiological Analysis

Samples of 25 g were transferred aseptically into individual stomacher bags (Seward Medical, UK), containing 225 ml of sterile Buffered Peptone Water (BPW) solution (0.1%) and homogenized in stomacher (Lab Blender 400, Seward Medical, UK) for 60 s. For each sample, further serial decimal dilutions were prepared in BPW solution (0.1%). For TVC determination, the amount of 0.1 ml of these serial dilutions of meat homogenates was spread on duplicate plates of Plate Count Agar (PCA, Merck, Darmstadt, Germany) and incubated at 30°C for 48 hr. *Pseudomonas* spp. were enumerated on *Pseudomonas* Agar Base (Oxoid, UK) supplemented with cetrimide, fucidine and cephaloridine (CFC) supplements (SR, 103, Basingstone, UK) providing a selective isolation medium for *Pseudomonas* spp. Colonies were counted after 48 hr at 30°C and the performance of an oxidase test. The LAB were enumerated by the pour plating technique on de Man Rogosa Sharpe Agar (MRS, Oxoid code CM361, Basingstone, UK) after incubation at 37°C for 3 days. Enterobacteriaceae were enumerated by the pour plating technique on Violet Red Bile Glucose Agar (Oxoid, UK) after incubation at 37°C for 24 hr. Finally, *Clostridium* spp. was enumerated using Reinforst Clostridium Medium (RCM, Merck code 1.05410) after incubation at 35°C for 2 days under anaerobic conditions. Anaerobic conditions were achieved by the use of Anaeropack® GENbox Jar combined with Pack-Anaero oxygen absorbers. Microbiological counts were expressed as log cfu/g.

Sensory Evaluation

After sampling, all meat samples were frozen at -30°C until sensory evaluation. Meat samples (ca. 100 g) were cooked in a microwave oven at high power (700 W) for 3-4 min including time of defrosting. A panel of 51 untrained judges, graduate students and faculty of the Department of Chemistry was used for sensory analysis (acceptability study). Panelists were asked to evaluate taste and odor of cooked samples. Along with the test samples, the panelists were presented with a freshly thawed meat sample, stored at -30°C throughout the experiment, this serving as the reference sample. Acceptability of odor and taste was estimated using an acceptability scale ranging from 5 to 0, with 5 corresponding to the most liked sample and 0 corresponding to the least liked sample. A score of 3.0 was taken as the lower limit of acceptability (Chouliara et al., 2007).

Physicochemical Analysis

The pH value was recorded using a Metrohm, model 691, pH meter. Meat samples (10 g) were thoroughly homogenized with 10 ml of distilled water and the homogenate used for pH determination. The TBA was determined according to the method proposed by Kirk and Sawyer (1991). The TBA content was expressed as mg of malondialdehyde (MDA)/kg meat.

Statistical Analysis

Experiments were replicated twice on different occasions with different ground meat samples. Analyses were run in triplicate for each replicate (n = 2 x 3). Microbiological data were transformed into logarithms of the number of colony forming units (cfu/g) and were subjected to analysis of variance (ANOVA). Means and standard deviations were calculated and when F-values were significant at the $P < 0.05$ level, mean differences were separated by the Least Significance Difference (LSD) procedure (Steel and Torrie, 1980).

DISCUSSION AND RESULTS

Microbiological Changes

The TVC values for all different meat treatments during storage are given in Figure 1. The initial TVC (day 0) value for the fresh ground meat was ca. 4.3 log cfu/g, indicative of good quality meat (Dawson et al., 1995). The TVC exceeded the value of 7 log cfu/g, considered as the upper microbiological limit for fresh meat, as defined by the ICMSF (1986), on day 4 for air packaged samples, day 5 for samples treated with chitosan 1% w/w, day 6 for samples packaged with the oxygen absorber and day 9 for samples containing both chitosan (1% w/w) and oxygen absorber. Thus a microbiological shelf life extension of 1 day was achieved for samples treated with chitosan, 2 days for samples packaged with the oxygen absorber alone and 5 days for samples treated with chitosan plus the oxygen absorber.

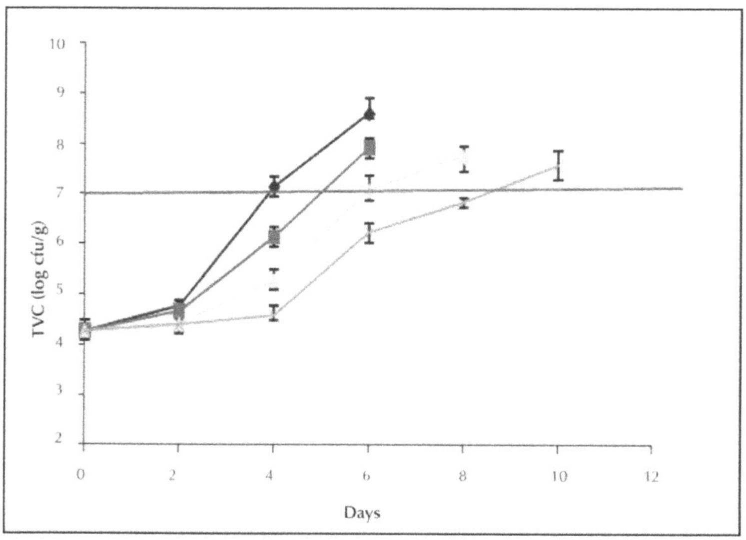

Figure 1. Combined effect of oxygen absorber and chitosan on TVC in minced meat stored at 4°C. (♦) control (air packaged), (■) chitosan 1% w/w, (▲) oxygen absorber, (×) chitosan 1% w/w + oxygen absorber.

In recent studies, Giatrakou et al. (2010) reported shelf life extension of 2 days in ready to eat chicken products stored at 4°C after the addition of chitosan (1.5% v/w). In turn, Sagoo et al. (2002) reported shelf life extension of 2 days in minced pork mixture after the addition of chitosan (0.6% v/w).

Present results are also in general agreement with those of Soultos et al. (2008) who reported a reduction of 0.5 and 1 log cfu/g in TVC of fresh pork sausages after 1 day of storage at 4°C with the addition of chitosan at concentration of 0.5 and 1%, respectively. Georgantelis et al. (2007) also reported that the addition of chitosan at a concentration of 10g/kg (1%) reduced the TVC of pork sausages by 2 log cfu/g after 10 days of storage at 4°C.

Regarding the use of the oxygen absorber, present results are in agreement with those of Martinez et al. (2006), who reported that samples stored in the absence of oxygen (either under vacuum or in the presence of an oxygen absorber) showed low TVC values which never reached the limit of 7 log cfu/g even after 20 days of storage.

The *Pseudomonads* (Figure 2) are strictly aerobic, gram-negative bacteria, one of the main spoilage microorganisms in fresh meat (Jay et al., 2005). Chitosan reduced the *Pseudomonads*' population by 0.6 log cfu/g on day 6 of storage ($P < 0.05$). As expected, under aerobic conditions, *Pseudomonads* formed a significant part of the microbial flora associated with the spoilage of meat. Use of the oxygen absorber resulted to complete inhibition of the *Pseudomonads* on day 2 of storage. The *Pseudomonads*' population remained under the method detection limit for 6 and 8 days respectively when using the oxygen absorber or the oxygen absorber plus chitosan combination. This is expected given that the *Pseudomonads* are strictly aerobic bacteria and are unable to grow in the absence of oxygen. Based on these findings, an additive effect of the oxygen absorber plus chitosan combination was shown on growth inhibition of the *Pseudomonads*.

Giatrakou et al. (2010) reported an increasing trend for *Pseudomonads* in control samples of ready to eat poultry products ranging from 5.1 to 9.5 log cfu/g, whereas respective counts for samples treated with chitosan were ca. 0.5 log cfu/g lower during 12 day storage period.

The same authors also investigated the combined effect of MAP and 1.5% chitosan on *Pseudomonads*' population and reported a reduction of 1 to 2.3 log cfu/g for samples packaged either under MAP alone or samples treated with chitosan and packaged under MAP, respectively. Georgantelis et al. (2007) reported reduction of 0.8 and 1.1 log cfu/g in *Pseudomonads*' population in pork sausages stored at 4°C after 5 and 10 days respectively with the addition of 1% chitosan.

Present results are also in general agreement with those of Soultos et al. (2008) who reported that the use of chitosan at a concentration of 0.5 and 1% (w/w) resulted to a reduction of 0.5 and 0.9 log cfu/g in *Pseudomonads*' population in pork sausages stored at 4°C.

Regarding the use of the oxygen absorber, Martinez et al. (2006) reported reduction in psychrotrophic counts by 1.7 log cfu/g in pork sausages stored at 4°C as compared to samples stored aerobically. Sheridan et al. (1997) also noted that meat packaged under vacuum had slower aerobic bacterial growth than meat packaged in modified atmospheres with oxygen. It is well documented that packaging in environments without oxygen retards aerobic microbial growth and delays spoilage due to slow proliferation of bacteria capable of tolerating anaerobic conditions (Martinez et al., 2006).

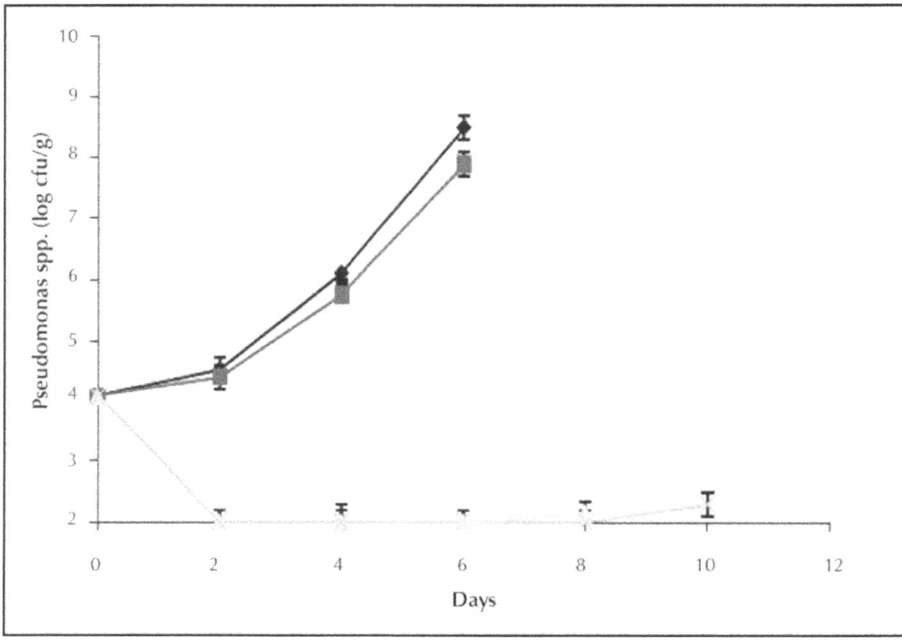

Figure 2. Combined effect of oxygen absorber and chitosan on *Pseudomonas* spp. populations in minced meat stored at 4°C. (♦) control (air packaged), (■) chitosan 1% w/w, (▲) oxygen absorber, (×) chitosan 1% w/w + oxygen absorber.

With respect to the Enterobacteriaceae (gram-negative, facultative anaerobes), considered to be a hygiene indicator (Zeitoun et al., 1994), their initial counts were 2.3 log cfu/g (Figure 3), indicative of good manufacturing practices in the processing plant. On day 6 of storage Enterobacteriaceae counts reached 3.9 log cfu/g for aerobically packaged samples. On the same day, counts were reduced by 0.7 log cfu/g in the chitosan or oxygen absorber treated samples and 1.7 log cfu/g in the chitosan plus oxygen absorber treated samples ($P < 0.05$).

Present results are in general agreement with those of Giatrakou et al. (2010), who reported that the use of chitosan (1.5% v/w) reduced Enterobacteriaceae counts from ca. 5.0 to ca. 4.0 log cfu/g in ready to eat chicken products after 6 days of storage at 4°C. Ouattara et al. (1997) reported that Enterobacteriaceae found in meat products (bologna, beef pastrami and cooked ham) were inhibited in the presence of antimicrobial films containing chitosan and acetic acid. Finally, Soultos et al. (2008) reported a reduction in Enterobacteriaceae population of 1.1 log cfu/g after the addition of chitosan (1% w/w) in pork sausages after 7 days of storage at 4°C.

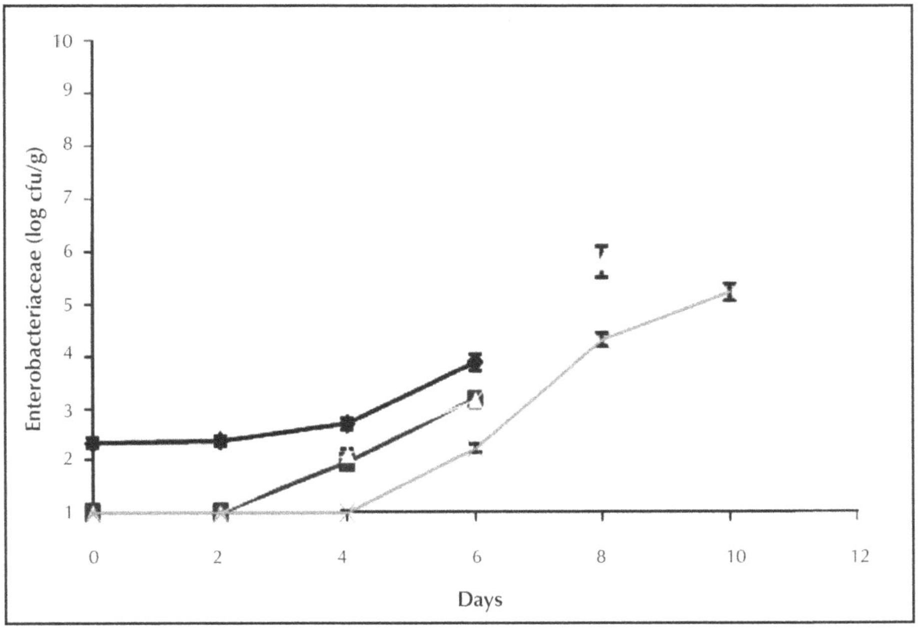

Figure 3. Combined effect of oxygen absorber and chitosan on Enterobacteriaceae populations in minced meat stored at 4°C. (♦) control (air packaged), (■) chitosan 1% w/w, (▲) oxygen absorber, (×) chitosan 1% w/w + oxygen absorber.

The LAB are gram-positive fermentative bacteria that can grow under different packaging conditions (in the presence or the absence of oxygen) and thus constitute a substantial part of the natural microflora of meats (Jay et al., 2005). The initial LAB counts (Figure 4) were ca. 2.8 log cfu/g (day 0) and reached 5.6 log cfu/g on day 6 of storage for the air packaged samples. Use of chitosan alone resulted in a slight reduction of LAB by 0.2 log cfu/g on day 6 of storage ($P > 0.05$), while the oxygen absorber, as expected, had no effect on LAB counts. On the same day, the combined use of chitosan (1% w/w) and oxygen absorber resulted in a reduction in LAB counts by 1.4 log cfu/g ($P < 0.05$).

These findings, with regard to the use of chitosan, are in general agreement with those of Soultos et al. (2007), who reported LAB reduction of 1.7 log cfu/g after the addition of chitosan (1% w/w) in pork sausages after 7 days of storage at 4°C. Giatrakou et al. (2010) also reported LAB reduction of 1 log cfu/g in ready to eat poultry products after the addition of chitosan (1% v/w) on day 6 of storage at 4°C. Samples containing chitosan showed lower microbial counts compared to control samples regardless of their Gram-negative or positive classification. The mode of antimicrobial action of chitosan may be through its interaction with the cell membrane and cell components, resulting in increased permeability of the membrane and leakage of cell material or due to its water-binding capacity and inhibition of various enzymes (Helander et al., 2001).

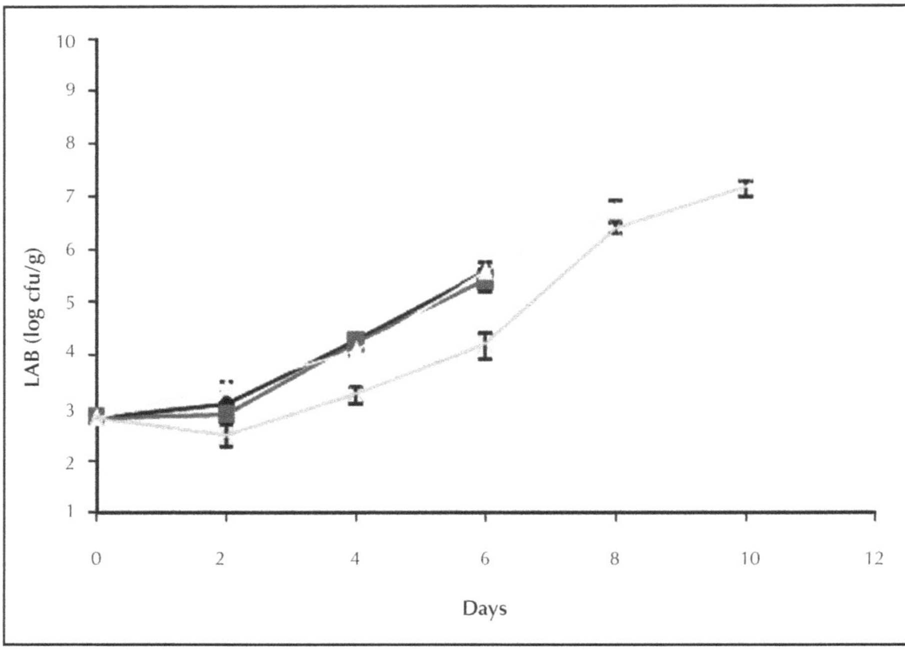

Figure 4. Combined effect of oxygen absorber and chitosan on LAB populations in minced meat stored at 4°C. (♦) control (air packaged), (■) chitosan 1% w/w, (▲) oxygen absorber, (×) chitosan 1% w/w + oxygen absorber.

Clostridium spp. counts remained below the method detection limit (1 log cfu/g) throughout storage for all samples.

Sensory Changes

Sensory properties (odor and taste) of cooked ground meat are given in Figures 5 and 6, respectively. The lower acceptability limit of 3.0 was reached for odor after ca. 2-3 days, for the air packaged samples, 3-4 days for samples treated with chitosan 1%, 6 days for samples packaged with the oxygen absorber and 8-9 days for samples packaged with the oxygen absorber plus chitosan. Respective values for taste were 3, 3-4, 6, and 9 days. Both odor and taste proved to be equally sensitive sensory attributes for product quality evaluation. It should be noted that as microbiological data also show, the samples packaged with the oxygen absorber as well as those treated with chitosan and packaged with the oxygen absorber always gave higher sensory scores ($P < 0.05$) as compared to controls and chitosan treated samples.

Giatrakou et al. (2010) reported a shelf life extension of 4 days in ready to eat chicken samples containing chitosan (1.5% v/w) as compared to the control samples.

Sensory data were in reasonably good agreement with microbiological data (TVC). The small differences observed between the two may be attributed to the fact that it is not the total number of microorganisms but rather the number of specific spoilage organisms (SSO) that are responsible for product deterioration (Jay et al., 2005).

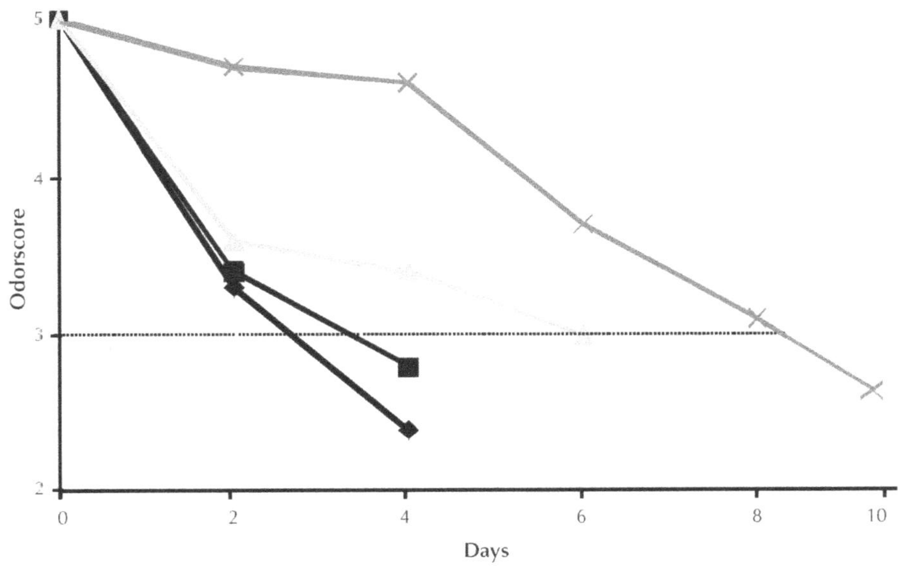

Figure 5. Combined effect of oxygen absorber and chitosan on odor minced meat stored at 4°C. (♦) control (air packaged), (■) chitosan 1% w/w, (▲) oxygen absorber, (×) chitosan 1% w/w + oxygen absorber.

Physicochemical Changes

Changes in pH for the different treatments, during the 10 day storage period, were statistically insignificant ($P > 0.05$). The pH values varied between 5.9 (day 0) and 6.0 (day 10) for aerobically packaged samples, 5.9 and 6.0 for chitosan treated samples and between 5.9 and 5.8 for samples packaged with the oxygen absorber and samples treated with chitosan and packaged with the oxygen absorber (data not shown).

The TBA values for different treatments are given in Figure 7. The TBA values varied between 0.06 and 2.9 mg MDA/kg meat. Initial MDA value of meat was 0.9 mg/kg. At the point of product sensory rejection, the MDA content of control samples was equal to 1.5 mg MDA/kg. Respective MDA values were 1 mg/kg for the chitosan treated samples and 0.4–0.5 mg/kg for samples packaged with the oxygen absorber plus chitosan. Thus, both chitosan and especially the oxygen absorber prevented lipid oxidation of ground meat.

These values are lower than the proposed by Buyn (2002) limit of 2 mg MDA/kg above which, rancid off-flavors become sensorily detectible in meat products. The above results are in agreement with those of Georgantelis et al. (2007), who reported a reduction in TBA values for pork sausages by 73% for samples containing 1% chitosan (0.25 mg MDA/kg) as compared to control samples (0.96 mg MDA/kg after 10 days of storage at 4°C. Darmadji and Izumimoto (1994) reported that TBA values of beef containing 1% chitosan remained unchanged after 10 days of storage at 4°C (0.5 mg MDA/kg), whereas respective values of control samples increased sharply.

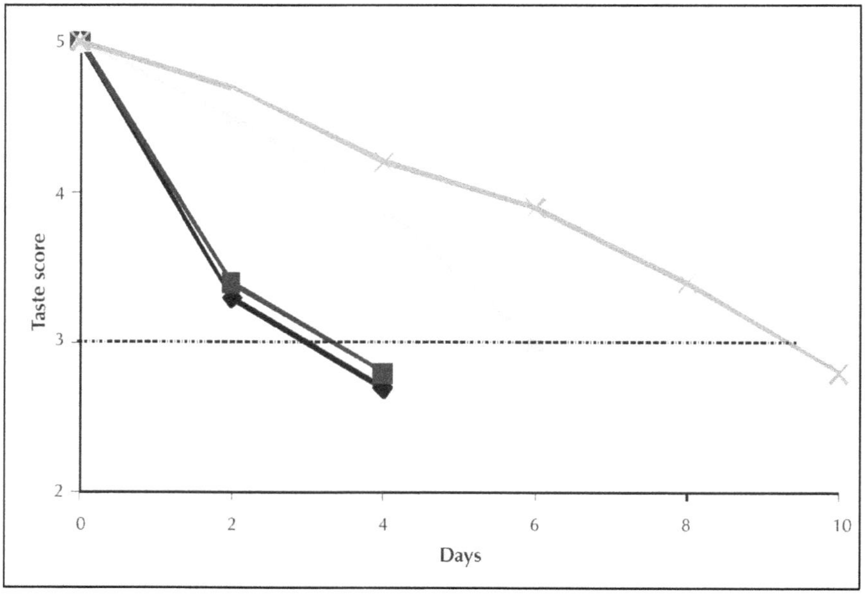

Figure 6. Combined effect of oxygen absorber and chitosan on taste minced meat stored at 4°C. (♦) control (air packaged), (■) chitosan 1% w/w, (▲) oxygen absorber, (×) chitosan 1% w/w + oxygen absorber.

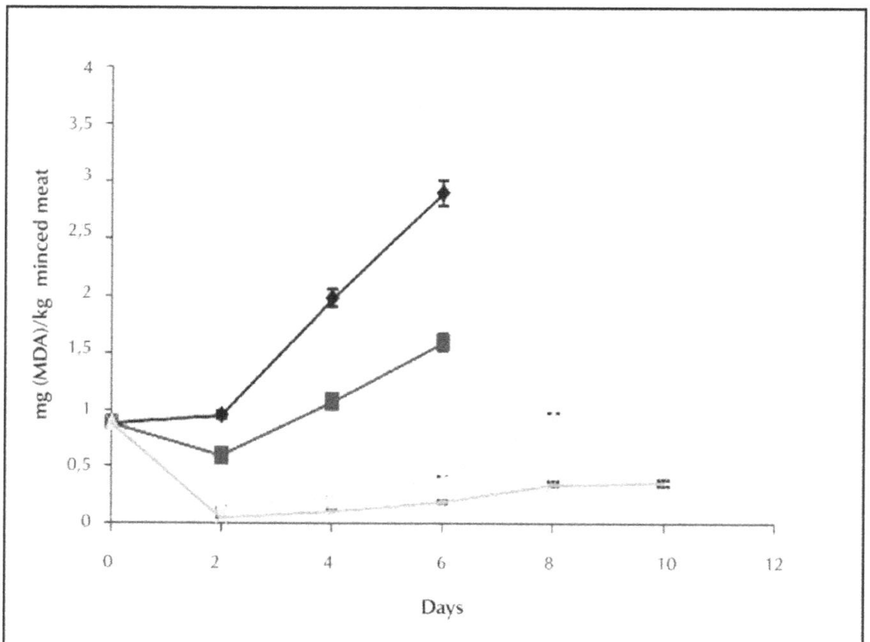

Figure 7. Combined effect of oxygen absorber and chitosan on degree of lipid oxidation in minced meat stored at 4°C. (♦) control (air packaged), (■) chitosan 1% w/w, (▲) oxygen absorber, (×) chitosan 1% w/w + oxygen absorber.

Regarding the use of the oxygen absorber, Martinez et al. (2006) reported a reduction in TBA values (0.5 mg MDA/kg) for pork sausages by 70% after 16 days of storage at 2°C compared to samples stored under MAP (80%O_2/20%N_2) (1.8 mg MDA/kg). Giatrakou et al. (2010) studied the combined effect of chitosan and vacuum packaging in chicken pepper kebab and reported a reduction in TBA values (1.0 mg MDA/kg) by 55% compared to control samples after 8 days at 4°C (2.3 mg MDA/kg).

The mechanism by which chitosan retards oxidative rancidity is thought to be related to chelation of free iron which is released from hemoproteins during storage (Yen et al., 2008).

CONCLUSION

Based both on microbiological and sensory data, it can be concluded that shelf life of ground beef samples was ca. 3 days for the air packaged samples, 3-4 samples for samples treated with 1% (v/w) chitosan, 6 days for samples packaged with oxygen absorber and 9 days for samples treated with chitosan and packaged with oxygen absorber. Thus chitosan and the oxygen absorber showed synergistic effect in shelf life extension of ground beef meat.

KEYWORDS

- **Chitosan**
- **Ground meat**
- **Oxygen absorber**
- **Shelf life extension**

ACKNOWLEDGMENTS

The authors would like to thank Mr. K. Yoshizaki (Oxygen Absorber Division), Mitsubishi Gas Chemical Company Inc. for providing the oxygen absorbers.

Chapter 10

Control of *E. coli* O157:H7 in Stirred Yogurt using Mastic and Almond Essential Oils

E. Pagiataki. E. Chouliara, and M. G. Kontominas

INTRODUCTION

Growth and survival of *Escherichia coli* O157:H7 inoculated (10^4 cfu/ml) into heat treated cow's milk pre- and post-fermentation, in the presence of mastic oil (0.05 and 0.1% v/v) and almond oil (1 and 2% v/v) was investigated. Yogurt was prepared with the addition of a 2% (v/v) mixed culture of *Lactobacillus bulgaricus* and *Streptococcus thermophilus* and incubation at 45°C for 4 hr. Fermented yogurt samples were subsequently stored at 4°C for 10 days. After 24 hr post-fermentation the pH decreased from 6.9 to 4.1 stabilizing at this value until the end of storage. Starter culture LAB was not substantially affected by the presence of essential oils at the concentrations used. No Yeasts and Moulds were detected in the product throughout storage. In the case of pre-fermentation inoculation of *E.coli* O157:H7, the pathogen was unable to grow at the fermentation temperature of 45°C. When inoculated post-fermentation into yogurt, *E.coli* O157:H7 survived for approximately 8 days in control samples, 2 days with the addition of mastic oil (0.05 or 0.1%) or almond oil (2%) and 6 days with the addition of almond oil (1%). Thus, even though mastic and almond oil have a controlling effect on *E.coli* O157:H7 they do not completely eliminate the pathogen in stirred yogurt during the first few days of refrigerated storage.

Yogurt is the product of milk fermentation with the addition of starter cultures of *Streptococcus thermophilus* and *Lactobacillus delbrueckii subsp. bulgaricus*. However, in some countries, like Australia, several other lactic acid bacteria have been used from time to time for the production of yogurt, such as *Lactobacillus helveticus* and *Lactobacillus jugurti* (Varnam and Sutherland, 1996). Fermentation is achieved after an initial pH drop between 4.6 and 4.7, while further substantial acidification is hampered by refrigeration at temperatures of 12–15°C in the case of stirred yogurt and at 2–5°C in the case of set yogurt.

The conditions, under which fermented milk products are produced, constitute a selective environment, which favors the development of yeasts and fungi as contaminant microorganisms. Yeasts are the most common microorganisms related to fermentations and leading to gas production. Genders such as *Kluyveromyces* and *Saccharomyces* occur more often in contaminated yogurt which contains added sugars, usually in the form of fruit pulps (Varnam and Sutherland, 1996). Fungi grow in the yogurt/air interface and lead to formation of spots on the product surface. A large variety of fungi have been isolated from yogurt, including *Aspergillus, Penicillium, Phizopus.* Besides fungi, bacteria such as *E.coli* may contaminate yogurt. Specifically serotype O157:H7 has been incriminated for several dairy product intoxications (Ogwaro et al., 2002). It has been documented that *E. coli* has the ability to adjust and is particularly tolerant to acid conditions prevailing in yogurt, thus managing to survive (Ryu and Beuchat, 1998).

Essential oils are complex mixtures of organic compounds that vary widely depending on the plant they originate from, plant variety, geographical origin and so on. The ingredients of essential oils can be classified into two groups: oxygen-containing and non-oxygen-containing. Non-oxygen-containing compounds include aliphatic and aromatic hydrocarbons, cyclic terpenes ($C_{10}H_{16}$), sesquiterpenes ($C_{15}H_{24}$), diterpenes and azulenes. Oxygen-containing compounds include higher hydrocarbon derivatives such as alcohols, aldehydes, ketones, organic acids, lactones, furan products, and phenols. Phenols are products of aromatic hydrocarbons, which derive through substitution of one or more hydrogen atoms on the aromatic ring by hydroxyl groups. Such phenolic compounds posses well documented antimicrobial properties (Burt, 2004; Cosentino et al., 1999,). Most studies on the effect of essential oils against spoilage as well as pathogenic microorganisms have concluded that essential oils are more effective against gram-positive versus gram-negative bacteria, which is justified by the presence of an external layer of lipopolysaccharides that surrounds the cell wall of gram-negative bacteria hampering the diffusion of hydrophobic compounds into the cell (Harpaz et al., 2003; Lambert et al., 2001). However, other studies report no difference in behavior between gram-positive and gram-negative bacteria to essential oils (Burt, 2004).

In the present study, two essential oils were used: mastic oil and almond oil because of their antibacterial activity and their organoleptic characteristics, compatible with yogurt. Mastic is a white, semitransparent, natural resin that is obtained as a trunk exudate from mastic trees. The mastic tree *Pistacia lentiscus*, is an evergreen bush that grows only in the southern part of the Greek island of Chios. Mastic gum has numerous uses including medicine that is mastic gum has been used in clinical trials on patients with peptic ulcers (Al-Habbal et al., 1984). In surgery, byproducts of mastic gum are used for the production of special sutures which are eventually absorbed by the human body. In dentistry, mastic acts as an oral antiseptic and tightens the gums (Topitsoglou-Themeli et al., 1984) being used in toothpastes and chewing gums. The essential oil of mastic gum is also used in perfumery and in the cosmetic industry for the production of creams and other facial products (Doukas, 2003). Moreover, there are numerous culinary uses of mastic gum, including biscuits, ice cream, and confectionery products.

Escherichia coli is relatively resistant to α-pinene, which is the most abundant compound (65%) of mastic oil (Jeon et al., 2001). *Escherichia coli* is also relatively resistant to β-myrcene, the second most abundant compound (25%) of mastic oil. The *p*-Cymene, β-caryophyllene(1%), methyl isoeugenol, limonene(1.5%), γ-terpinene, and *trans*-anethole show only moderate antibacterial activity, and in some cases most bacteria are resistant to them (Koutsoudaki et al., 2005). Verbenone (0.07%), α-terpineol (0.01%), and linalool (0.5%) are some of the trace components of mastic oil, showing higher antibacterial activity than all its other components.

Almond oil (*Oleum amygdalae*) has long been used in complementary medicine circles for its numerous health benefits. Almonds and the oil they produce, have many properties including anti-inflammatory, immunity-boosting, and anti-hepatotoxicity effects (Al-Said et al.1986; Huwez 1998, 1986; Iauk et al., 1996). The oil is often applied to the skin as an emollient and has been traditionally used as a skin lubricant. It is mild, lightweight oil that can be used in a variety of applications, as a substitute for olive oil (http://en.wikipedia.org/wiki/Almond). The almond tree is native to hot climates (Doukas, 2003; Magiatis et al., 1999) Almonds are a good source of antioxidants (http://www.raysahelian.com/almond.html) including the flavonoids: isorhamnetin-3-*O*-rutinoside ,isorhamnetin-3-*O*-glucoside, catechin, kaempferol-3-*O*-rutinoside, epicatechin, quercetin-3-*O*-galactoside, and isorhamnetin-3-*O*-galactoside. Among almond oil's constituents α-tocopherol contributes to its antimicrobial activity (http://chimikoergastirio.blogspot.com).

Based on the above, the objective of the present study was to investigate the effect of both mastic and almond oil in controlling the growth of *E.coli* O157:H7 inoculated in (a) milk prior to yogurt fermentation and (b) yogurt immediately after fermentation.

MATERIALS AND METHODS

Materials

In the present study, cow's milk (3.5% fat) was used for the preparation of yogurt. Yogurt culture 4C-350 consisting of *Streptococcus thermophilus* and *Lactobacillus delbrueckii subsp. bulgaricus* was obtained from Chr. Hansen A/S, Hørsholm, Denmark. Mastic oil was supplied by KOKKINAKIS S.A. (Athens, Greece). Almond oil was supplied BIOLIFE S.A. (Crete, Greece). Mastic oil was used at concentrations 0.05% and 0.1% (v/v). Almond oil was used at concentrations 1% and 2% (v/v). From preliminary sensory tests it was found that almond oil could be used at substantially higher concentrations than mastic oil without adverse sensory effects.

A non virulent *E.coli* O157:H7 strain was donated by Prof. J. Farkas from the stock collection of the Svent Isvan University in Budapest, Hungary. The microorganism was maintained on Trypticase soy agar (Difco Lab. Detroit MC, USA) slants and subcultured monthly.

Yogurt Production

The 3 l of milk were placed in 5 l stainless container and heated to 85°C for approximately 10 min. Subsequently, milk was dispensed into sterile plastic cups of 500 ml capacity. Cooling of the milk at ambient temperature was continued until its temperature reached ca. 45°C. at which time, it was inoculated with the yogurt culture at a rate of 2.0%. Cups were transferred to a constant temperature (45°C) incubator and left to solidify until pH reached a value of 4.5-4.6. Yogurt was then transferred to ambient temperature for ca. 30 min. Finally, the product was placed in the refrigerator at 4°C. Two different experiments were conducted regarding the stage of *E. coli* O157:H7 inoculation.

Experiment I

In the first experiment *E. coli* O157:H7, at a concentration of 10^4 cfu/ml, was added into the heat treated milk right after the addition of the yogurt culture and the procedure of yogurt preparation was followed as previously described.

Subsequently, after yoghurt preparation (day 0), almond oil or mastic oil was added to the product using a pipette under continuous stirring. Five experimental treatments resulted as follows: treatment 1 (I0) no essential oil was added (control); treatment 2 (AI1) almond oil at a concentration of 1% was added; treatment 3 (AI2) almond oil at a concentration of 2% was added; treatment 4 (MI1) mastic oil at a concentration of 0.05% was added and treatment 5 (MI2) mastic oil at a concentration of 0.1% (v/w) was added.

Experiment II

In the second experiment *Escherichia coli* O157:H7, at a concentration of 10^4 cfu/ml, was inoculated into the solidified yogurt under continuous stirring. The yogurt was then placed in the refrigerator at 4°C as previously described.

Subsequently, almond oil or mastic oil was added as described in experiment 1. Five experimental treatments resulted as follows: treatment 1 (II0) no essential oil was added (control); treatment 2 (AII1) almond oil at a concentration of 1% was added; treatment 3 (AII2) almond oil at a concentration of 2% was added; treatment 4 (MII1) mastic oil at a concentration of 0.05% was added, and treatment 5 (MII2) mastic oil at a concentration of 0.1% was added. The first experiment simulated the contamination of heat treated milk (raw material), while the second simulated the contamination of the end product with *E. coli* O157:H7.

Microbiological Analysis

The following groups of microorganisms were determined: Total Viable Count (TVC), Yeasts-Moulds, Lactic Acid Bacteria, and *E. coli* O157:H7. An amount of 10g from the yogurt prepared was transferred to Stomacher bag (Seward Medical, London, UK) containing 90 ml of peptone water solution 0.1% and homogenized for 30 s using the Lab Blender stomacher (Seward Medical, London, UK).

The TVC was determined on Plate Count Agar (PCA, OXOID, CM 325), after incubation at 30°C for 48 hr. Yeasts-Moulds were enumerated on Rose Bengal Chloramphenicol agar (RBC, OXOID CM 549) after incubation at 30°C for 5 days. Lactic Acid Bacteria were enumerated after mixing 1 ml of the sample with approximately 15 ml of Man Rogosa Sharpe agar (MRS, OXOID CM 361). After solidification of the first agar layer, a second layer was formed (approximately 10 ml of the respective agar) and plates were incubated at 37°C for 3 days. Finally, for the enumeration of *Escherichia coli* O157:H7, Sorbitol MacConkey Agar (SMAC, OXOID CM) was used and incubation was carried out at 37°C for 3 days.

Determination of pH

The pH was determined using the method of AOAC (1995).

Statistical Treatment

Experiments were replicated twice on different occasions with different yogurt samples. Analysis were run in triplicate for each replicate (n = 2 x 3). Microbiological data were transformed into logarithms of the number of colony forming units (cfu/g) and were subjected to analysis of variance using the software SPSS 16 for windows. Means and standard deviations were calculated and when F values were significant at the ($p < 0.05$) level.

DISCUSSION AND RESULTS

Experiment I

Changes in TVC of substrate inoculated with *E.coli* O157:H7 prior to fermentation with the addition of either mastic oil or almond oil are given in Figures 1(a) and (b) respectively. As shown in Figure 1(a), TVC on day 0 (ca. 6 hr after the addition of mastic oil) was 7.8 log cfu/g for the control samples (no essential oil addition), 7.2 log cfu/g for samples with 0.05% v/v of mastic oil and 7.1 log cfu/g for samples with with 0.1% v/v of mastic oil. Thus, mastic oil resulted to a statistically significant ($p < 0.05$) reduction of 0.6-0.7 log cfu/g in product TVC on the same day of its addition. Mastic oil concentration, at least within the concentration range tested, did not affect TVC.

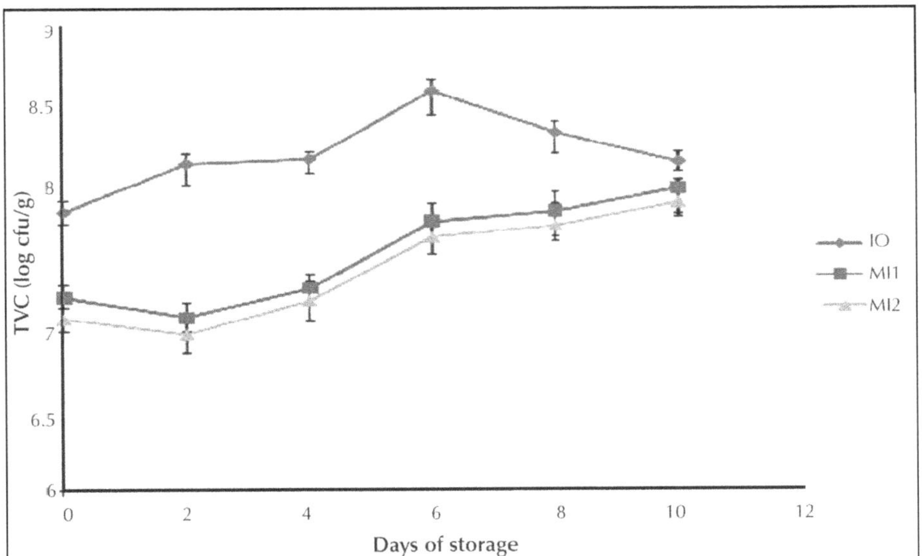

Figure 1(a). Changes in TVC of substrate inoculated pre-fermentation with *E.coli* O157:H7 with the addition of mastic oil during storage at 4°C (—◆— IO *E.coli* O157:H7-control; —■— MI1 *E.coli* O157:H7 plus mastic oil 0.05%/v/v; —▲— MI2 *E.coli* O157:H7 plus mastic oil 0.1%/v/v).

The TVC of samples containing mastic oil goes through an adaptation period due to the stress applied by the essential oil and subsequently increases to 7.9-8.0 log cfu/g on day 10 of storage. What is interesting to note, is that the TVC of control samples reached the same concentration (8.1 log cfu/g) on day 10 of storage implying complete recovery of mesophilic microflora with respect to the stress applied by the essential oil.

Analogous is the effect of almond oil on TVC (Figure 1b). Initial TVC of control samples was 7.8 log cfu/g increasing to 8.1 log cfu/g on day 10 of storage. Addition of 1 and 2% almond oil resulted to a reduction of TVC by 0.50 and 0.55 log cfu/g ($p < 0.05$) respectively. In the case of almond oil an increase in its concentration produced small but statistically significant differences ($p < 0.05$) in TVC beginning with day 2 of storage. At the end of storage TVC of all three treatments was between 7.9 and 8.1 log cfu/g.

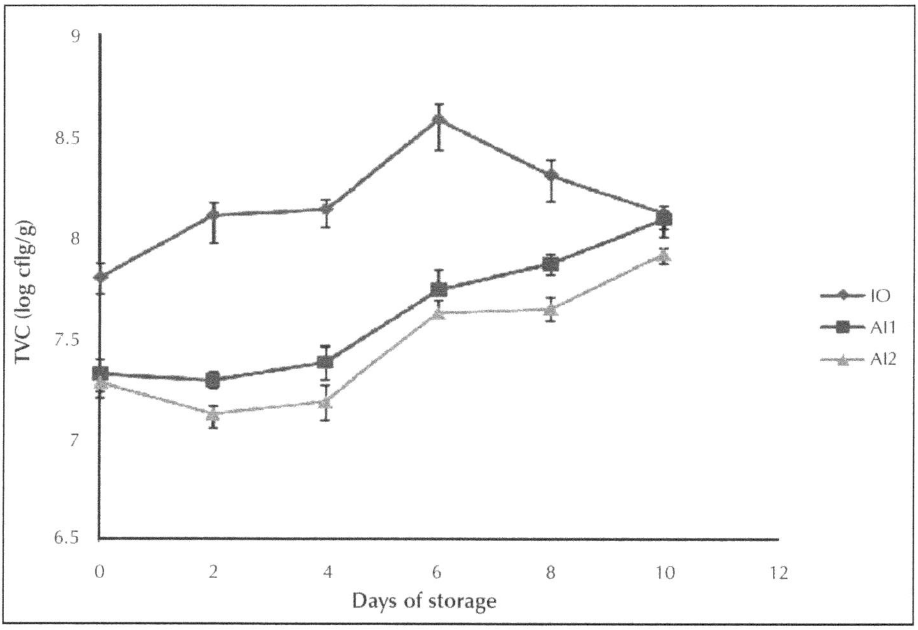

Figure 1(b). Changes in TVC of substrate inoculated pre-fermentation with *E.coli* O157:H7 with the addition of almond oil during storage at 4°C (—◆— IO *E.coli* O157:H7-control; —■— AI1 *E.coli* O157:H7 plus almond oil 1%/v/v; —▲— AI2 *E.coli* O157:H7 plus almond oil 2%/v/v).

Analogous is the effect of almond oil on TVC (Figure 1b). Initial TVC of control samples was 7.8 log cfu/g increasing to 8.1 log cfu/g on day 10 of storage. Addition of 1 and 2% almond oil resulted to a reduction of TVC by 0.50 and 0.55 log cfu/g ($p < 0.05$) respectively. In the case of almond oil an increase in its concentration produced small but statistically significant differences ($p < 0.05$) in TVC beginning with day 2 of storage. At the end of storage TVC of all three treatments was between 7.9 and 8.1 log cfu/g.

Based on the above, it is clear that both mastic and almond oils have a small but statistically significant ($p < 0.05$) antimicrobial effect on TVC of yogurt inoculated with *E.coli* O157:H7. Of the two essential oils mastic oil proved to be substantially more effective attaining an equivalent reduction in TVC at only 1:20 the concentration of almond oil. Such a difference in antimicrobial activity between the two essential oils may be related to α-pinene, the major compound of mastic oil (65% w/w) exhibiting strong antimicrobial properties (Koutsoudaki et al., 2005). Most essential oils have been classified as GRAS and have been used for both the extension of shelf life of a variety of foods as well as the inhibition of numerous pathogens including *L. monocytogenes*, *E.coli* O157:H7, *Staphylococcus aureus*, *Salmonella* spp. and so on (Burt, 2004).

Present results on TVC are in general agreement with those of Otaibi et al. (2008) regarding "Labneh" (strained yogurt) consumed in the Middle East after the addition of sage, thyme and marjoram oil at concentrations equal to 0.2, 0.5, and 1 ppm. Increase in concentration of essential oil employed, resulted in reduction of product TVC.

The TVC increased reaching a maximum on day 7 of storage (ca. 8.1 log cfu/g) and subsequently decreased until day 10 of storage. Likewise, Penney et al. (2004) reported that vanillin at a concentration of 2000 ppm suppressed fungal and total microbial growth in yogurt to which wild blueberries had been added. On the other hand, the addition of fruit flavors (cherries, oranges, strawberries and bananas) did not have a significant effect on total bacterial populations (Con et al., 1996).

Changes in LAB populations of substrate inoculated with *E.coli* O157:H7 prior to fermentation with the addition of either mastic or almond oil are given in Figures 2(a) and (b), respectively.

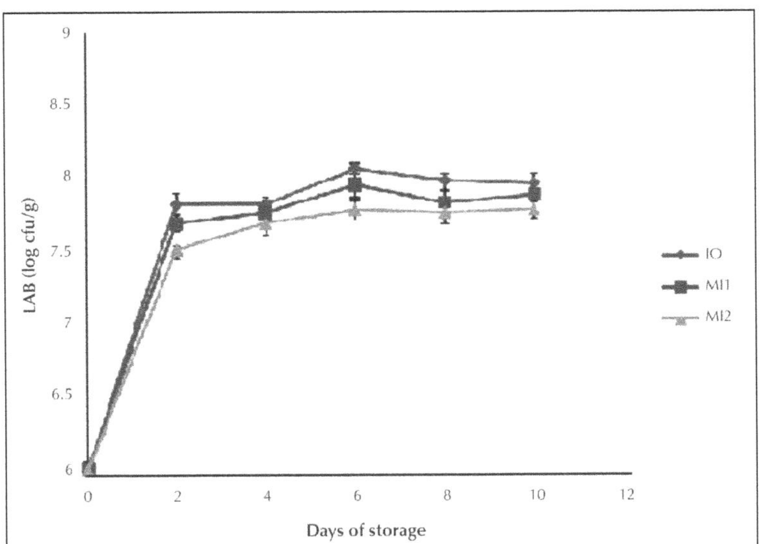

Figure 2(a). Changes in LAB populations of substrate inoculated pre fermentation with *E.coli* O157:H7 with the addition of mastic oil during storage at 4°C (—◆— IO *E.coli* O157:H7-control; —■— MI1 *E.coli* O157:H7 plus mastic oil 0.05%/v/v; —✱— MI2 *E.coli* O157:H7 plus mastic oil 0.1%/v/v).

As shown in Figure 2(a), LAB count on day 0 was 6 log cfu/g for all three treatments. On day 2 of storage LAB count rose to 7.8, 7.7, and 7.5 log cfu/g for control samples, samples containing 0.05 and 0.1% of mastic oil respectively. On day 10 the LAB count for all treatments was 7.8–8.0 log cfu/g respectively. It is thus clear that mastic oil at the lower concentration (0.05%) did not affect starter culture bacteria *Lactobacillus bulgaricus* and *Streptococcus thermophilus* growth. The LAB is gram-positive fermentative bacteria which are resistant to essential oils (Burt, 2004). A small but statistically significant ($p < 0.05$) difference in LAB count was observed between the control samples and samples treated with 0.1% mastic oil. This is in general agreement with Vinderola et al. (2002) who showed that Lactic acid starter and probiotic (bifidobacteria) bacterial growth was not affected by flavorings such as vanilla, banana, and strawberry added to pure cultures at concentrations equal to those used in commercial practice. An effect was shown only at higher concentrations of the above flavorings tested.

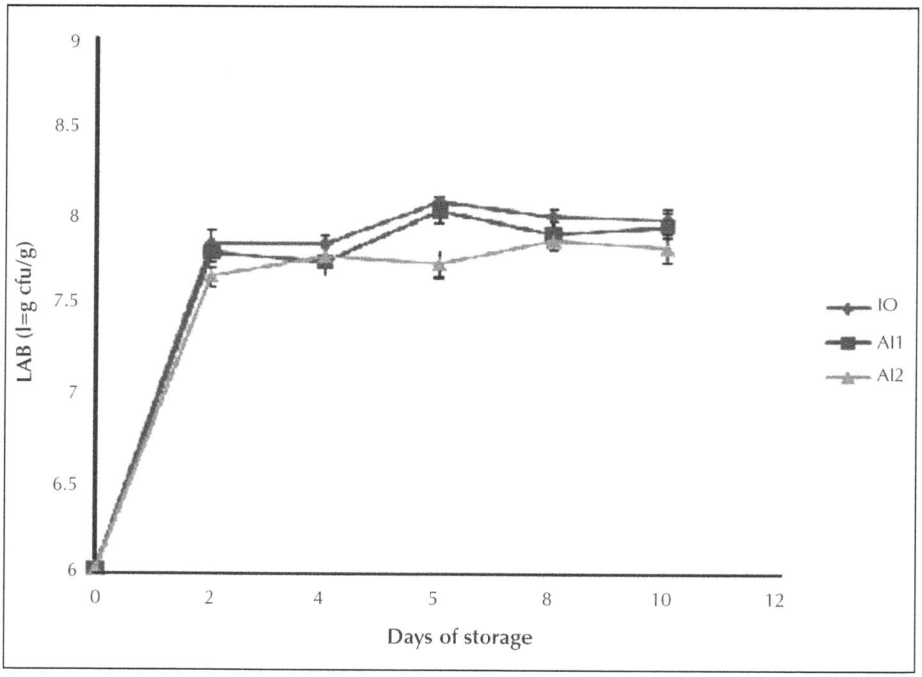

Figure 2(b). Changes in LAB populations of substrate inoculated pre fermentation with *E.coli* O157:H7 with the addition of almond oil during storage at 4°C (—◆— IO *E.coli* O157:H7-control; —■— AI1 *E.coli* O157:H7 plus almond oil 1%/v/v; —▲— MI2 *E.coli* O157:H7 plus almond oil 2%/v/v).

After the addition of thyme, marjoram and sage essential oils to strained yogurt (Labneh), Otaibi and Demerdash (2008) reported that *Streptococcus thermophilus* and *Lactobacillus bulgaricus* were not inhibited by low concentrations of the essential oils. Notably, El-Nawawy et al. (1998) reported that the presence of certain herbs, including thyme in the manufacture of yogurt, increased counts of *S. thermophilus* and *L. bulgaricus* compared to untreated controls during storage. Likewise, Jaziri et al. (2009) reported no effect of green and black tea added at levels 2 and 4% (w/v) to yogurt on starter culture populations (9.0–8.5 log cfu/ml) over a 6 week storage period at 4°C. Legislation in several EU countries (Anonymous, 2001) requires the presence of at least 10^7 starter culture bacteria per g of yogurt at the time of product consumption. The LAB populations in the present study (7.7–8.1 log cfu/g) exceeded this value after day 1 of storage. An identical pattern for LAB was obtained in the case of yogurt containing 1 and 2% almond oil.

With regard to *E.coli* O157:H7 inoculated into milk prior to fermentation, no growth occurred throughout storage. This may be attributed to (a) the high temperature of fermentation; according to the literature *E.coli* O157:H7 will grow poorly or not at all at temperatures 44–45.5°C (Desmarchelier and Grau, 1997), (b) the low pH (<4.4) attained during fermentation; *E.coli* will grow at a pH between 4.4 and 10. The use of starter cultures for rapid acidification contributes significantly to the control of pathogens. The type of acidulant used also influences *E.coli* growth, that is *E.coli* O157:H7 will grow at 37°C in a medium of pH 4.5 acidified with hydrochloric acid but not with lactic acid. Despite the above, *E.coli* have several different regulatory systems that enable cells to adapt and survive acid stress. Furthermore, combinations of parameters such as temperature, pH, aw, essential oils, salts and so on act as hurdles for controlling the growth of *E.coli* in dairy products (Desmarchelier and Grau, 1997). In addition to the above two factors, competition between the dominant LAB and *E.coli* O157:H7 may have also contributed to growth inhibition of *E.coli* O157:H7.

Based on present results it is clear that *E.coli* O157:H7 will not survive the conventional yogurt manufacturing process if introduced as a contaminant prior to fermentation.

Experiment II

Changes on TVC of yogurt inoculated with *E.coli* O157:H7 post fermentation with the addition of either mastic or almond oil is given in Figures 3(a) and (b) respectively. As shown in Figure 3(a) TVC on day 0 (ca. 6 hr after the addition of mastic oil) was 7.6 log cfu/g for control samples (no essential oil added), 7.1 log cfu/g for samples with 0.05% mastic oil and 7.0 log cfu/g for samples with 0.1% mastic oil. Thus, mastic oil resulted to a statistically significant ($p < 0.05$) reduction of 0.5-0.6 log cfu/g in TVC on the same day of its addition. In contrast to experiment I, concentration of mastic oil (0.05% vs 0.1%) affected ($p < 0.05$) TVC of yogurt at least until day 6-7 of storage.

The TVC of control samples increased from an initial 7.6 to 8.3 log cfu/g on day 10 of storage; respective TVC for samples containing 0.05% mastic oil were 7.1, rising to 7.9 log cfu/g and 7.0 to 7.8 log cfu/g for samples containing 0.1% mastic oil. Overall, from the beginning to the end of storage, mastic oil resulted to the reduction of TVC by ca. 0.5-0.6 log cfu/g as compared to control samples exhibiting a mild bacteriocidal effect ($p < 0.05$).

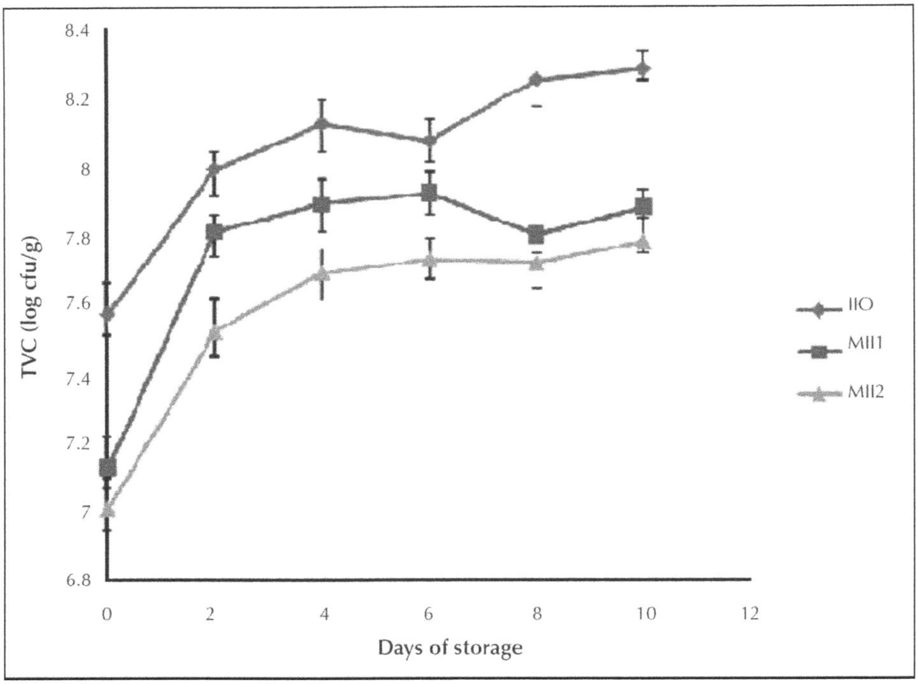

Figure 3(a). Changes in TVC of yogurt inoculated post fermentation with *E.coli* O157:H7 with the addition of mastic oil during storage at 4°C (IIO *E.coli* O157:H7-control; MII1 *E.coli* O157:H7 plus mastic oil 0.05% /v/v; MII2 *E.coli* O157:H7 plus mastic oil 0.1%/v/v).

Analogous was the effect of almond oil on TVC (Figure 3(b)). Initial TVC of control samples was 7.6 log cfu/g increasing to 8.3 log cfu/g on day 10 of storage. Addition of 1 and 2% almond oil resulted to the reduction of TVC by 0.35 and 0.50 log cfu/g respectively ($p < 0.05$). Concentration of almond oil did not seem to affect TVC at least within the concentration range tested. Overall from the beginning to the end of storage almond oil resulted to a reduction of TVC by ca. 0.4 log cfu/g ($p < 0.05$) as compared to control samples.

Based on the above data it is clear that both essential oils had a mild antimicrobial effect ($p < 0.05$) on TVC of yogurt inoculated with *E.coli* O157:H7. As in the first experiment, mastic oil had a similar antimicrobial effect to almond at only a fraction (1:20) of its concentration.

Changes in LAB populations of yogurt inoculated with *E.coli* O157:H7 post-fermentation with the addition of either mastic or almond oil are given in Figures 4(a) and (b) respectively. As shown in Figure 4(a), the LAB count on day 0 was 6 log cfu/g for all three treatments. On day 2 of storage the LAB count reached 7.80, 7.65, and 7.5 log cfu/g for control samples and samples containing 0.05 and 0.1% of mastic oil respectively. On day 10 of storage the LAB count for all treatments was 7.7-7.9 log cfu/g. It is clear that mastic oil, at the lower concentration used, did not affect ($p > 0.05$) the starter culture bacteria. A small but significant ($p < 0.05$) difference in LAB count was observed between the control samples and samples treated with 0.1% mastic oil.

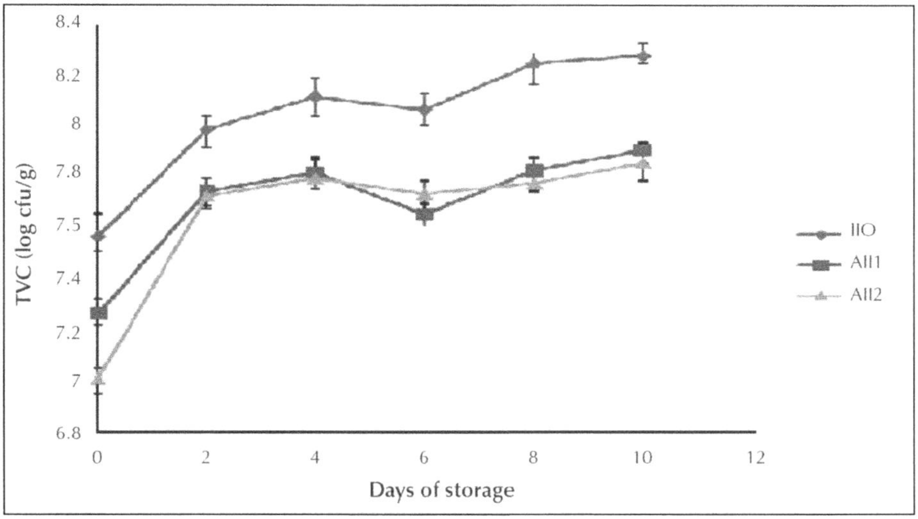

Figure 3(b). Changes in TVC of yogurt inoculated post fermentation with *E.coli* O157:H7 with the addition of almond oil during storage at 4°C (—◆— IIO *E.coli* O157:H7-control; —■— AII1 *E.coli* O157:H7 plus almond oil 1% /v/v; —▲— AII2 *E.coli* O157:H7 plus almond oil 2%/v/v).

A similar pattern for LAB was obtained in the case of yogurt containing 1 and 2% almond oil. Here, essential oil concentration had no effect ($p > 0.05$) on LAB populations and only after day 4 of storage a significant difference ($p < 0.05$) was noted between controls and samples containing the essential oil. The LAB populations recorded in the present study are in agreement with those of Hamann and Marth (1984) reporting LAB populations in excess of 10^6 cfu/g in yogurt.

Changes in *E.coli* O157:H7 inoculated into yogurt post-fermentation with the addition of either mastic or almond oil is given in Figures 5(a) and (b) respectively. As shown in Figure 5(a), *E.coli* O157:H7 population on day 0 was 4 log cfu/g for control samples, 3.4 log cfu/g for samples containing 0.05% mastic oil and 3.1 log cfu/g for samples containing 0.1% mastic oil. Thus, in only a few hours after the addition of the essential oil a reduction equal to 0.6 and 0.9 log cfu/g in *E.coli* O157:H7 was achieved ($p < 0.05$) for a concentration of 0.05 and 0.1% of mastic oil respectively.

On day 2 of storage *E.coli* O157:H7 was totally inhibited in samples containing mastic oil irrespective of its concentration. On the other hand, it took ca. 8 days for total inhibition of *E.coli* O157:H7 in control samples. In the latter case pH reduction from an initial value of 6.9 to a final value of 4.1 (Table 1) led to a constant reduction in *E.coli* O157:H7 population until its total inhibition. In the case of addition of mastic oil, *E.coli* O157:H7 was under the combined effect of three different hurdles: those of low pH, essential oil and competition from LAB leading to its complete inhibition in only 2 days (Desmarchelier and Grau, 1997).

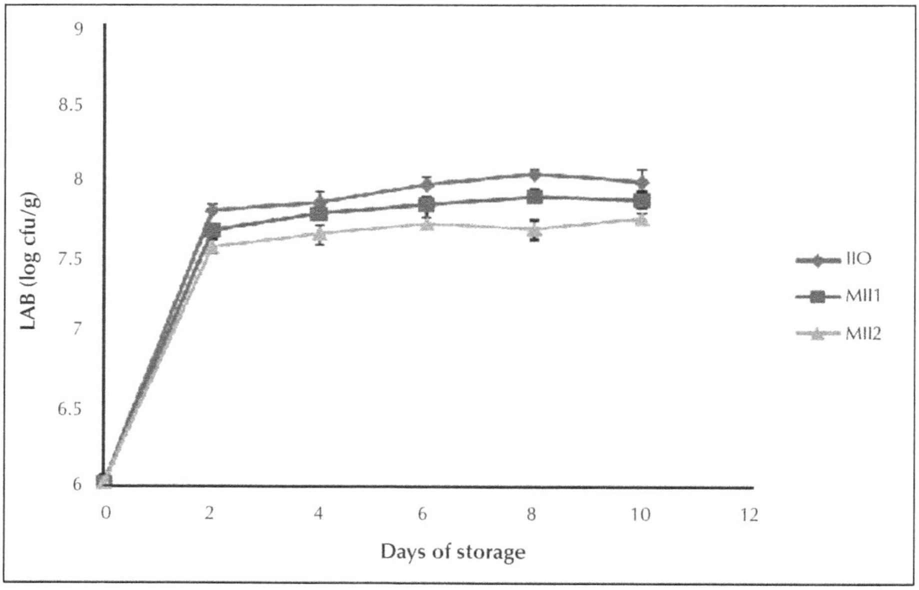

Figure 4(a). Changes in LAB populations of yogurt inoculated post fermentation with *E.coli* O157:H7 with the addition of mastic oil during storage at 4°C (IIO *E.coli* O157:H7-control; MII1 *E.coli* O157:H7 plus mastic oil 0.05%/v/v; MII2 *E.coli* O157:H7 plus mastic oil 0.1%/v/v).

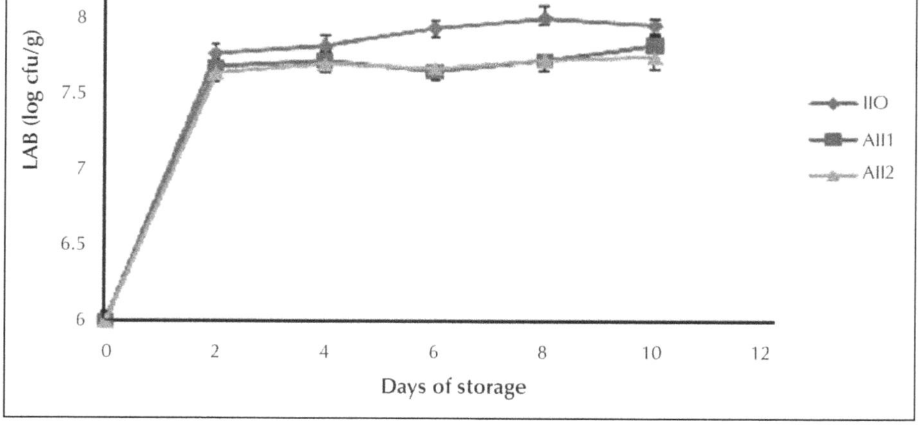

Figure 4(b). Changes in LAB populations of yogurt inoculated post fermentation with *E.coli* O157:H7 with the addition of almond oil during storage at 4°C (IIO *E.coli* O157:H7-control; AII1 *E.coli* O157:H7 plus almond oil 1%/v/v; AII2 *E.coli* O157:H7 plus almond oil 2%/v/v).

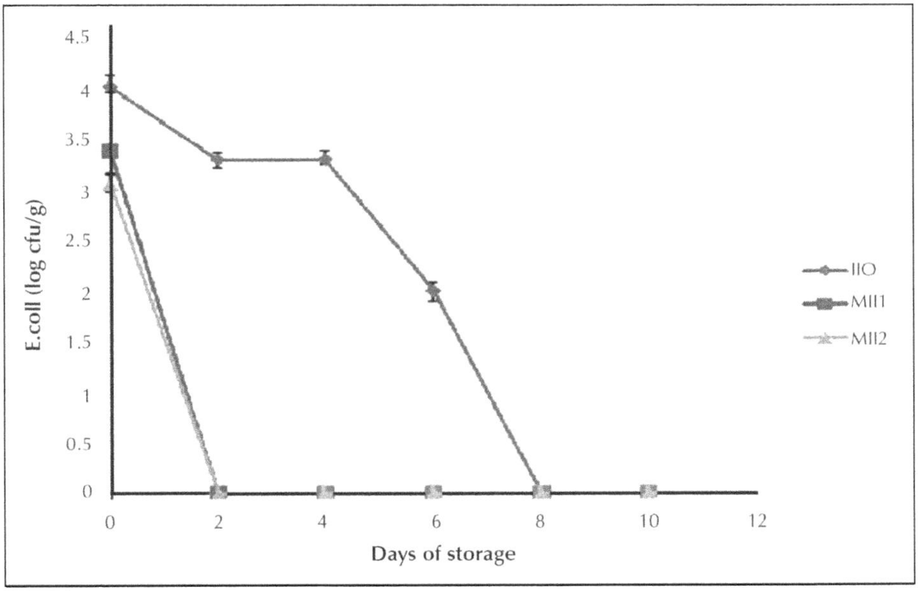

Figure 5(a). Changes in *E.coli* O157:H7 populations of yogurt inoculated post fermentation (10^4 cfu/g) into yogurt with the addition of mastic oil during storage at 4°C (IIO *E.coli* O157:H7-control; MII1 *E.coli* O157:H7 plus mastic oil 0.05%/v/v; MII2 *E.coli* O157:H7 plus mastic oil 0.1%/v/v).

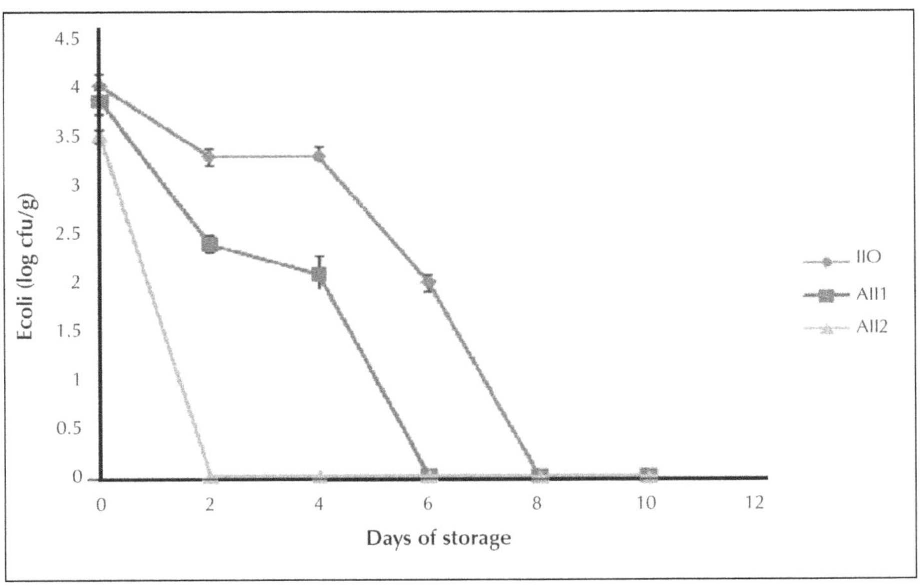

Figure 5(b). Changes in *E.coli* O157:H7 populations of yogurt inoculated post fermentation (10^4 cfu/g) into yogurt with the addition of almond oil during storage at 4°C (IIO *E.coli* O157:H7-control; AII1 *E.coli* O157:H7 plus almond oil 1%/v/v; AII2 *E.coli* O157:H7 plus almond oil 2%/v/v).

Although acidic foods such as yogurt may be considered as safe due to their low pH, *E.coli* O157:H7 has been implicated in food poisoning resulting from the consumption of yogurt (Morgan et al., 1993). Apple cider (Besser et al., 1993) and mayonnaise (Neagan et al., 1994) have also been implicated in similar food poisoning episodes.

Table 1. Changes in pH during fermentation of milk and storage of yogurt produced at 4°C.

Time (h)	pH value
0	6.85
6	6.24
12	5.02
18	4.44
24	4.11
48	4.10
72	4.12
96	4.08
120	4.10
144	4.05
168	4.07
192	4.12
216	4.10
240	4.09

It has also been reported that *E.coli* O157:H7 inoculated into yogurt may remain viable 1–8 days under refrigeration (Massas et al., 1997). Leyer et al. (1995) showed that *E.coli* O157:H7 has specific mechanisms of adaptation to acidic environment which are favored by the presence of lactic acid. Despite this fact, data in Figure 5(a) show that besides low pH and competition from large populations of LAB, mastic oil substantially affected *E.coli* O157:H7 growth.

Hsin-Yi and Chou (2001) reported that acid adaptation of *E.coli* O157:H7 at 7°C enhanced its growth in asparagus and mangoes' juice. In contrast, *E.coli* O157:H7 populations (both in natural and acid adapted strains) were completely inhibited after 5 and 4 days of storage respectively at 7°C.

Penney et al. (2004) inoculated acid adapted *E.coli* at 10^3 and 10^6 cfu/g and incubated the product at 4°C. After 14 hr there was no detectable survival in yogurt receiving 10^3 cfu/g; when the inoculums was increased to 10^6 cfu/g only 0.02% of the bacterium survived.

Koutsoudaki et al. (2005) showed that *E.coli* was relatively resistant to α-pinene, the most abundant component of mastic oil; it was also resistant to β-myrcene, the second most abundant component of mastic oil and to β-pinene. According to these authors the compounds responsible for mastic oil antibacterial activity include: verbenone, linalool and α-terpineol.

Ogwaro et al. (2002) inoculated *E.coli* O157:H7 into yogurt both prior to and post fermentation (10^5 cfu/g). In the first case *E.coli* O157:H7 was totally inhibited after 4 days while in the second case a 1 log cfu/g reduction in the bacterium population was recorded after 6 days of incubation at 4°C. These results are in general agreement with those reported in the present study in which the population of *E.coli* O157:H7 was reduced by ca. 2 log cfu/g after 6 days (control samples). In agreement to our findings, Ogwaro et al. (2002) concluded that *E.coli* O157:H7 cells inoculated prior to fermentation declined more rapidly than cells inoculated post-fermentation, despite the reduction in pH being similar. This may be explained by the stress applied to cells by the high fermentation temperature, being unable to recover when maintained in an acidic environment. Canganella et al. (1998) inoculated (10^2–10^3 and 10^4–10^6 cfu/g) fruit yogurt with several microorganisms including *E.coli*. When stored at 4°C, *E.coli* decreased to undetectable levels after 24d (low inoculum) and 28d (high inoculum). It was shown by these authors that the Lactobacillus strain was more effective compared to the Streptococcus strain in inhibiting *E.coli* in liquid cultures. It was concluded that the presence of these species in high numbers certainly represents a biological barrier against the survival of potential contaminants such as *E.coli* (Tin et al., 1996).

Errendilek (2007) inoculated *E.coli* O157:H7 into plain yogurt (10^8 cfu/g) and stored the product at 4°C for 35 days. He reported that *E.coli* O157:H7 decreased sharply to 3 log cfu/g after only 1 day of storage and reached undetectable levels after 10 days, which is in reasonable agreement to our results (control samples).

Other studies revealed that *E.coli* O157:H7 survived in acidic conditions in Yakult, a diluted fermented milk, for 4 hr (pH = 3.6) (Hsin-Yi and Chou (2001) and in acidophilus yogurts for 48h (pH = 4.60–4.96) (Kasimoglou and Akgus, 2004).

With respect to the effect of plant extracts and essential oils on *E.coli* O157:H7 survival, Kotzekidou et al. (2008) showed that among the four extracts (apricot, lemon, plum and strawberry) added at a concentration of 0.1 ml/100 g of chocolate, lemon was the most inhibitory to *E.coli* O157:H7. At low temperature storage (7°C) a nearly 1.7 log cfu/g reduction in *E.coli* O157:H7 was observed in samples inoculated with ca. 10^5 cfu/g and treated with lemon. A quick repression of the inoculum was observed in 2 days. Strawberry was able to repress *E.coli* O157:H7 by 1 log cfu/g during 7 days of storage at 7°C. In contrast, plum inhibited the growth of the inoculated *E.coli* O157:H7 showing bacteriostatic results and a reduction of 0.4 log cfu/g after 9 days of storage.

Kang and Fung (1999) reported that diacetyl, a major flavor constituent in fermented dairy products, at a concentration of 50 ppm, strongly inhibited the growth of *E.coli* O157:H7 inoculated into meat during fermentation.

Finally Schelz et al. (2006) reported that both sage and thyme essential oils had a clear inhibitory effect against *E.coli*. Gram-negative bacteria possess an outer membrane hydrophilic in nature composed of lipopolysactheride molecules (Nikaido, 1996). Small hydrophilic solutes are able to penetrate the outer membrane through pores in the membrane, providing hydrophilic trans-membrane channels, whereas the outer membrane serves as a penetration barrier to macromolecules and to hydrophobic compounds (Nikaido, 1996). Despite this fact, the outer membrane is not totally impermeable to hydrophobic molecules, some of which can slowly penetrate the cell wall through these pores. By passing the outer membrane is a prerequisite for any solute to exert bacteriocidal activity.

Plant extracts and essential oils may exhibit different modes of action against bacterial strains such as interference with the phospholipid bilayer of the cell membrane which results in increased permeability and loss of cellular constituents, damage of the enzymes involved in the production of cellular energy and synthesis of structural components and destruction or inactivation of genetic material (Kim et al., 1997). In general, the mechanism of action is considered to be the disturbance of the cytoplasmic membrane, disrupting the proton motive force electron flow, active transport, and coagulation of cell contents (Davidson, 1997; Sikkema et al., 1995).

Results regarding the effect of almond oil on growth of *E.coli* O157:H7 (Figure 5(b)) were somewhat different than those for mastic oil. At the higher concentration (2%) almond oil completely inhibited *E.coli* O157:H7 in 2 days, similar to mastic oil (0.05%) but at the lower concentration (1%) almond oil was considerably less effective requiring ca. 6 days for complete inhibition of the pathogen. On the other hand, according to Kotzekidou et al. (2008) almond oil exhibits a clear antibacterial activity against *E.coli* O157:H7 in chocolate while mastic oil is ineffective in controlling the pathogen.

Based on the above data regarding *E.coli* O157:H7, it is clear that the point in time of inoculation of the pathogen in yogurt (pre- or post-fermentation) is crucial for its survival. Ogwaro et al. (2002) stored yogurt at 4°C for 6 days and reported that when *E.coli* O157:H7 (10^5 cfu/g) is inoculated prior to fermentation, the pathogen is undetectable in 4 days. When, on the other hand, inoculation takes place post-fermentation (43°C), *E.coli* O157:H7 is still detectable (4 log cfu/g) after 6 days of storage at 4°C. It is most probable that exposure of the pathogen to high temperatures (43°C) causes an additional stress making recovery of *E.coli* O157:H7 impossible in an acidic environment.

Similarly Lee and Chen (2005) reported that when *E.coli* O157:H7 is inoculated into stirred yogurt prior to fermentation (43°C), its population is reduced by 4–5 log cfu/g during the first 10 days of storage at 4°C. When, however, the inoculation takes place post-fermentation, its population is reduced less than 2 log cfu/g. These results are in general agreement to those of the present study given the differences in experimental design (i.e. use of essential oils).

Finally, Errendilek (2007) inoculated (7.97 log cfu/ml) a series of yogurt products (yogurt drink, plain yogurt and salted yogurt) with *E.coli* O157:H7 and monitored its growth for 35 days at 4 and 22°C. Results showed a small reduction of the pathogen's population at 22°C but a substantial one at 4°C. After 10 days of storage there was a 7 log cfu/g reduction in *E.coli* O157:H7 population reported.

Yeasts and moulds were not detected in both experiments I and II. Present results are in agreement with those Otaibi and Demerdash (2008) who reported absence of yeasts and moulds in the fermented milk product "Labneh" with the addition of sage, thyme and marjoram stored at 5°C for 21 days. Yeasts and moulds were detected only in control samples (absence of essential oils) only after 14 and 21 days of storage (0.31×10^3 and 0.41×10^3 cfu/g, respectively). Results are also in agreement with those of Schelz and Molhar-Hohaman. (2006) and Hassan et al. (2001) who reported that the essential oil from thyme has antifungal and antimicrobial activities.

In contrast, Errendilek (2007) inoculated plain yogurt, salted yogurt and yogurt drink with *E.coli* O157:H7 (10^6–10^7 cfu/ml) and reported a yeast and mould population of 2–6.3 log cfu/ml in the yogurt drink, 2.3–5.9 log cfu/ml in plain yogurt and 3.5–6.4 log cfu/ml in salted yogurt during storage for 35 days at 4°C.

Finally, Canganella et al. (1998) inoculated *E.coli* O157:H7, *Rhodotorula mucilaginosa* and *Kluyveromyces marxianus* (10^2–10^3 and 10^4-10^6 log cfu/ml) in fruit yogurt and reported a population of 10^2-10^6 cfu/ml for *Kluyveromyces marxianus* during storagefor 30 days at 4°C. The population of *Rhodotorula mucilaginosa* increased to 10^3-10^5 under the same storage conditions. Yeast growth was attributed to the fruit content of yogurt. Contamination by yeasts and moulds is one of the main limiting factors for the stability and commercial value of yogurt (Deak, 1991).

CONCLUSION

The present study shows that both mastic and almond oil, at the concentrations tested, do not substantially affect LAB (yogurt cultures). They contribute to the inhibition of yeasts and moulds imparting at the same time a desirable flavor to cow's yogurt. *E.coli* O157:H7 inoculated into heat treated milk pre-fermentation cannot grow at the fermentation temperature of 45°C. When inoculated into yogurt post-fermentation, *E.coli* O157:H7 survives for approximately 8 days in control samples, 2 days with the addition of mastic oil (0.05 or 0.1%) or almond oil (2%) and 6 days with the addition of almond oil (1%).

Eventhough both essential oils exhibit a clear antimicrobial effect against *E.coli* O157:H7, they do not eliminate potential health risks associated with post-fermentation contamination of yogurt by this pathogen. Thus, with regard to food safety, the post-fermentation process should be carefully inspected so as to eliminate possible contamination of yogurt by *E.coli* O157:H7.

KEYWORDS

- **Almond oil**
- **Mastic oil**
- **Pre-fermentation**
- **Post-fermentation inoculation**
- **Yogurt**

References

1

Bryant, E. A., Fulton, G. P., and Budd, G. L. (1992). *Disinfection alternatives for safe drinking water*. Van Nostrand Reinhold, New York, p. 518.

Carmody, E. P., James, K. J., and Kelly, S. S. (1996). Dinophysistoxin-2: The predominant diarrhoetic shellfish toxin in Ireland. *Toxicon* **34**, 351–359.

Christian, B. and Luckas, B. (2008). Determination of marine biotoxins relevant for regulations from the mousse bioassay to coupled LC-MS methods. *Analytical and Bioanalytical Chemistry* **391**, 117–134.

Croci, L., Toti, L., De Medici, D., and Cozzi, L. (1994). Diarrhetic shellfish poison in mussels: Comparison of methods of detection and determination of the effectiveness of depuration. *International Journal of Food Microbiology* **24**, 337–342.

EU Directive 86/609/EEC (1986). Council Directive 86/609/EEC of 24 November, 1986 on the approximation of laws, regulations and administrative provisions of the Member States regarding the protection of animals used for experimental and others scientific purposes. *Official Journal of European Community* **L358**, 1–28.

EU Regulation. 853/2004/EC (2004). Regulation (EC) No 853/2004 of the European Parliament and of the Council of 29 April, 2004 laying down specific hygiene rules for food of animal origin, L226 (25/06/2004), Brussels, pp. 22–82.

EU Regulation. 2074/2005/EC (2005). Commission Regulation (EC) No 2074/2005 of 5 December, 2005 laying down implementing measures for certain products under Regulation (EC) No 853/2004 of the European Parliament and of the Council and for the organization of official controls under Regulation (EC) No 854/2004 of the European Parliament and of the Council and Regulation (EC) No 882/2004 of the European Parliament and of the Council, derogating from Regulation (EC) No 852/2004 of the European Parliament and of the Council and amending Regulations (EC) No 853/2004 and (EC) No 854/2004, L338 (22/12/2005), Brussels, pp. 27–59.

EU Recommendation 2007/526/EC (2007). Commission Recommendation of 18 June, 2007 on guidelines for the accommodation and care of animals used for experimental and other scientific purposes. *Official Journal of European Community* **L197**, 1–89.

Fernandez, M. L., Richard, D. J. A., and Cembella, A. D. (2003). In *Manual on Harmful Marine Microalgae*. G. M. Hallegraeff, D. M.Anderson, and A. D. Cembella (Eds.). *In vivo* assays for phycotoxins, UNESCO Publishing, Paris, France, pp. 347–380.

Fux, E., McMillan, D., Bire, R., and Hess, P. (2007). Development of an ultra-performance liquid chromatography—mass spectrometry method for the detection of lipophilic marine toxins. *Journal of Chromatography A* **1157**, 273–280.

Gacutan, R. Q., Tabbu, M. Y., De Castro, T., Gallego, A. B., and Bulalacao, M. (1984). In *Proceedings of a Consultative Meeting Held in Singapore 11-14 September.* A. W. White, M. Anraku, and K. K. Hooi (Eds.). Detoxification of *Pyrodinium*-generated shellfish poisoning toxin in *Perna viridis* from Western Samar, Philippines. Southeast Asian Fisheries Development Center and the International Development Research Centre, pp. 80–85.

Gacutan, R. Q., Tabbu, M. Y., Aujero, E., and Icatlo, F. Jr. (1985). Paralytic shellfish poisoning due to *Pyrodinium bahamense* var. *compressa* in Mati, Davao Oriental, Philippines. *Marine Biology* **87**, 223–227.

Gago-Martinez, A., Rodriquez-Vazquez, J. A., Thibault, P., and Quilliam, M. A. (1996). Simultaneous occurrence of diarrhetic and paralytic shellfish poisoning toxins in Spanish mussels in 1993. *National Toxins* **4**, 72–79.

Gerssen, A., Mulder, P. P. J., McElhinney, M. A., and de, Boer. J. (2009). Liquid chromatography-tandem mass spectrometry method for the detection of marine lipophilic toxins under alkaline conditions. *Journal of Chromatography A* **1216**, 1421–1430.

Gerssen, A., Mulder, P., van, Rhijin. H., and de Boer, J. (2008). Mass spectrometric analysis of the marine lipophilic biotoxins pectenotoxin-2 and okadaic acid by four figgerent types of mass spectrometers. *Journal of Mass Spectrometry* **43**, 1140–1147.

Gonzalez, J. C., Fontal, O. I., Vieytes, M. R., Vieites, J. M., and Botana, L. M. (2002). Basis for a new procedure to eliminate diarrheic shellfish toxins from a contaminated matrix. *Journal of Agriculture and Food Chemistry* **50**, 400–405.

Gonzalez, J. C., Leira, F., Vieytes, M. R., Vieites, J. M., Botana, A. M., and Botana, L. M. (2000). Development and validation of a high-performance liquid chromatographic method using fluorimetric detection for the determination of the diarrhetic shellfish poisoning toxin okadaic acid without chlorinated solvents. *Journal of Chromatography A* **876**, 117–125.

Goto, H., Igarashi, T., Yamamoto, M., Yasuda, M., Sekiquchi, R., Watai, M., Tanno, K., and Yasumoto, T. (2001). Quantitative determination of marine toxins associated with diarrhetic shellfish poisoning by liquid chromatography coupled with mass spectrometry. *Journal of Chromatography A* **907**, 181–189.

Hamano, Y., Kinoshita, Y., and Yasumoto, T. (1986). Enteropathogenicity of diarrhetic shellfish toxins in intestinal models. *Journal of the Food Hygienic Society of Japan* **27**, 375–379.

Kelly, S. S., Bishop, A. G., Carmody, E. R., and James, K. J. (1996). Isolation of dinophysistoxin-2 and the high-performance liquid chromatography of diarrhetic shellfish toxins using derivatisation with 1-bromoacetylpyrene. *Journal of Chromatography A* **749**, 33–40.

Khadre, M. A., Yousef, A. E., and Kim, J. G. (2001). Microbiological aspects of ozone applications in food: A review. *Journal of Food Science* **6**, 1242–1252.

Koukaras, K. and Nikolaidis, G. (2004). Dinophysis blooms in Greek coastal waters (Thermaikos Gulf, NW Aegean Sea). *Journal of Plankton Research* **26**, 445–457.

Lawrence, J. F. and Scoot, P. M. (1993). In *Environmental Analysis*. D. Barcelo (Ed.). Techniques, applications and quality assurance, p. **273**.

Lee, J. S., Igarashi, T., Fraga, S., Dahl, E., Hovgaard, P., and Yasumoto, T. (1989). Determination of diarrhetic shellfish toxins in various dinoflagellate species. *Journal of Applied Phycology* **1**, 147–152.

Lee, J. S., Yanagi, T., Kenma, R., and Yasumoto, T. (1987). Fluorometric determination of diarrhetic shellfish toxins by high-performance liquid chromatography. *Agricultural and Biological Chemistry* **51**, 877–881.

Luckas, B. (1992). Phycotoxins in seafood-toxicological and chromatographic aspects. *Journal of Chromatography A* **624**, 439–456.

Mountfort, D. O., Kennedy, G., Garthwaite, I., Quilliam, M., Truman, P., and Hannah, D. J. (1999). Evaluation of fluorometric protein phosphate inhibition assay in the determination of okadaic acid in mussels. *Toxicon* **37**, 909–922.

Mouratidou, T., Kaniou-Grigoriadou, I., Samara, C., and Kouimtzis, T. (2004). Determination of okadaic acid and related toxins in Greek mussels by HPLC with fluorimetric detection. *Journal of Liquid Chromatography and Related Technologies* **27**, 2153–2166.

Mouratidou, T., Kaniou-Grigoriakou, I., Samara, C., and Kouimtzis, T. (2006). Detection of the marine toxin okadaic acid in mussels during a diarrhetic shellfish poisoning (DSP) episode in Thermaikos Gulf, Greece, using biological, chemical and immunological methods. *Science of the Total Environment* **366**, 894–904.

Murata, M., Shimatani, M., Sugitani, H., Oshima, Y., and Yasumoto, T. (1982). Isolation and structural elucidation of the causative toxin of the diarrhetic shellfish poisoning. *Bulletin of the Japanese Society for the Science of Fish* **48**, 549–552.

Pleasance, S., Quilliam, M. A., and Marr, J. C. (1992). Ionspray mass spectrometry of marine toxins. IV. Determination of diarrhetic shellfish poisoning toxins in mussel tissue by liquid chromatography/mass spectrometry. *Rapid Communications is Mass Spectroscopy* **6**, 121–127.

Prassopoulou, E., Katikou, P., Georgantelis, D., and Kyritsakis, A. (2009). Detection of okadaic acid and related esters in mussels during diarrhetic shellfish poisoning (DSP) episodes in Greece using the mousse bioassay, the PP2A inhibition assay and HPLC with fluorimetric detection. *Toxicon* **53**, 214–227.

Reboreda, A., Lago, J., Chapela, M. J., Vieites, J. M., Botana, L. M., Alfonso, A., and Cabado A. G. (2010). Decrease of marine toxin content in bivalves by industrial processes. *Toxicon* **55**, 235–243.

Reizopoulou, S., Strogyloudi, E., Giannakourou, A., Pagou, K., Hatzianestis, I., Pyrgaki, C., and Graneli, E. (2008). Okadaic acid accumulation in macrofilter feeders subjected to natural blooms of *Dinophysis* acuminate. *Harmful Algae* **7**, 228–234.

Rositano, J., Nicholson, B. C., and Pieronne, P. (1998). Destruction of cyanobacterial toxins by ozone. *Ozone-Science and Engineering* **20,** 209–215.

Schneider, K. and Rodrick, G. (1995). The use of ozone to degrade *Gymnodinium breve* toxins. 2. Conference Internationale sur la Purification des Coquillages, Rennes (France), 6–8 April, 1992, Actes de colloques, Ifremer, Brest, pp. 277–289, http://archimer.ifremer.fr/doc/00000/1621/.

Soriano, N. U., Migo, V. P., and Matsumura, M. (2003). Functional group analysis during ozonation of sunflower oil methyl esters by FT-IR and NMR. *Chemistry and Physics of Lipids* **126**, 133–140.

Smith, L. L. (2004). Oxygen, oxysterols, ouabain, and ozone: A cautionary tale. *Free Radical Biology and Medicine* **37**, 318–324.

Stover, E. L. and Jarnis, R. W. (1981). Obtaining high level wastewater disinfection with ozone. *Journal Water Pollution Control Federation* **53**, 1637–1647.

Suzuki, T., Jin, T., Shirota, Y., Mitsuya, T., Okumura, Y., and Kamiyama, T. (2005). Quantification of lipophilic toxins associated with diarrhetic shellfish poisoning in Japanese bivalves by liquid chromatography—mass spectrometry and comparison with mouse bioassay. *Fisheries Science* **71**, 1370–1378.

Suzuki, T., and Yasumoto, T. (2000). Liquid chromatography—electrospray ionization mass spectrometry of the diarrhetic shellfish—poisoning toxins okadaic acid, dinophysistoxin-1 and pectenotoxin-6 in bivalves. *Journal of Chromatography A* **874**, 199–206.

Suzuki, T., Yoshizawa, R., Kawamura, T., and Yamasaki, M. (1996). Interference of free fatty acids from the hepatopancreas of mussels with the mouse bioassay for shellfish toxins. *Lipids* **31**, 641–645.

Takagi, T. Hayashi, K., and Itabashi, Y. (1984). Toxic effects of free unsaturated fatty acids in the mouse bioassay of diarrhetic shellfish toxin by intraperitoneal injection. *Bulletin of the Japanese Society for the Science of Fish* **50**, 1413–1418.

Vale, P. and Sampayo, M. A. M. (1999). Esters of okadaic acid and dinophysistoxin-2 in Portuguese bivalves related to human poisonings. *Toxicon* **37**, 1109–1121.

Vale, P. and Sampayo, M. A. M. (2002). Esterification of DSP toxins by Portuguese bivalves form the Northwest coast dermined by LC-MS a widespread phenomenon. *Toxicon* **40**, 33–42.

Van Egmond, H. P., Aune, T., Lassus, P., Speijers, G. J. A., and Waldock, M. (1993). Paralytic and diarrhetic shellfish poisons: Occurrence in Europe, toxicity analysis and regulation. *Journal of Natural Toxins* **2**, 41–83.

White, A. W., Martin, J. L., Legresley, M., and Blogoslawski, W. J. (1985). In *Toxic Dinoflagellates.* D. M. Anderson, A. W. White, and D. G. Baden (Eds.). Inability of ozonation to detoxify shellfish toxins in the soft-shell clams. Elsevier Science Publishers, New York, pp. 473–478.

Yasumoto, T., Murata, M., and Ashima, Y. (1985). Diarrhetic shellfish toxins. *Tetrahedron* **41**, 1019–1025.

Yasumoto, T., Oshima, Y., Sugawara, W., Fukuyo, Y., Oguri, H., Igarishi, T., and Fujita, N. (1980). Identification of Dinophysis fortii as the causative organism of diarrhetic shellfish poisoning. *Bulletin of the Japanese Society for the Science of Fish* **46**, 1405–1411.

Yasumoto, T., Oshima, Y., and Yamaguchi, M. (1978). Occurrence of a new type shellfish poisoning in the Tohoku district. *Bulletin of the Japanese Society for the Science of Fish* **44**, 1249–1255.

2

Allen, J. C. and Joseph, G. (1985). Deterioration of pasteurized milk on storage. *Journal of Dairy Research* **52**, 469–487.

Bakish, R. and Hatfield, E. (1997). Coextrusions for flexible packaging. In *Wiley Encyclopedia of Packaging Technology.* A. L. Brody and K. S.

Marsch (Eds.). John Wiley and Sons, New York, pp. 237-240 and 629–638).

Barnard, S. E. (1972). Importance of shelf life for consumers of milk. *Journal of Dairy Science* **55**, 134–136.

Bosset, J. O., Gallman, P. U., and Sieber, R. (1994). Influence of light transmittance of packaging materials on the shelf-life of milk and dairy products—a review. In: *Food Packaging and Preservation*. M. Mathlouthi (Ed.). Chapman and Hall, London, pp. 222–263.

Cladman, W., Scheffer, S., Goodrich, N., and Griffiths, M. W. (1998). Shelf life of milk packaged in plastic containers with and without treatment to reduce light transmission. *International Dairy Journal* **8**, 629–636.

DeMan, J. M. (1978). Possibilities of prevention of light-induced quality loss of milk. *Canadian Institute of Food Science and Technology* **11**, 152–154.

DeMan, J. M. (1983). Light Induced Destruction of Vitamin A in Milk. *Journal of Dairy Science* **64**, 2031-3032.

Defosse, M. (2000). Novel milk bottle designs keep blow molders on top. *Modern Plastics Worldwide* **77**, 114–116.

Desarzens, C., Bosset, J. O., and Blanc, B. (1983). Light-induced changes of milk and some milk products. 1. Changes of color, taste and different vitamin contents. *Lebensmittel Wissenschaft und Technologie* **16**, 241–247.

Fanelli, A. J., Burlew, J. V., and Gabriel, M. K. (1985). Protection of milk packaged in high density polyethylene against photodegradation by fluorescent light. *Journal of Food Protection* **48**, 112–117.

Gaylord, A. M., Warthesen, J. J., and Smith, D. E. (1986). Influence of milk fat, milk solids, and light intensity on the light stability of vitamin A and riboflavin in low fat milk. *Journal of Dairy Science* **69**, 2779–2784.

Hoskin, J. C. (1988). Effect of fluorescent light on flavor and riboflavin content of milk held in modified half-gallon containers. *Journal of Food Protection* **51**, 19–23.

Hoskin, J. C. and Dimick, P.S. (1979). Evaluation of fluorescent light on flavor and riboflavin content of milk held in gallon returnable containers. *Journal of Food Protection* **42**, 105–109.

IDF, 99A. (1987). Sensory evaluation of dairy products.

Marsili, R. T. (1999). Comparison of SPME and dynamic headspace method for the GC-MS analysis of light-induced lipid oxidation products in milk. *Journal of Chromatographic Science* **37**, 17–23.

Moyssiadi, T., Badeka, A., Kondyli, E., Vakirtzi, T., Savvaidis, I., and Kontominas, M. G. (2004). Effect of light transmittance and oxygen permeability of various packaging materials on keeping quality of low fat pasteurized milk: Chemical and sensorial aspects. *International Dairy Journal* **14**, 429–436.

Papachristou, C., Badeka, A., Chouliara, E., Kondyli, E., Athanasoulas, A., and Kontominas, M. G. (2006a). Evaluation of polyethylene terphthalate as a packaging material for premium quality whole pasteurized milk in Greece. *European Food Research and Technology* **223**, 711–718.

Papachristou, C., Badeka, A., Chouliara, E., Kondyli, E., Kourtis, L., and Kontominas, M. G. (2006b). Evaluation of polyethylene terephthalate as a packaging material for premium quality whole pasteurized milk in Greece. *European Food Research and Technology* **224**, 237–247.

Rysstad, G., Ebbesey, A., and Eggestad, J. (1998). Sensory and chemical quality of UHT milk stored in paperboard cartons with different oxygen and light barriers. *Food Additives and Contaminants* **15**, 112–122.

Sattar, A., de Man, J. M., and Alexander, J. C. (1977). Wavelength effect on light-induced decomposition of vitamin A and beta-carotene in solutions and milk fat. *Canadian Institute of Food Science and Technology* **10**, 56–60.

Schröder, M. J. A. (1982). Effect of oxygen on keeping the quality of milk. *Journal of Dairy Research* **49**, 407–424.

Schröder, M. J. A., Scott, K. J., Bland, K. J., and Bishop, D. R. (1985). Flavor and vitamin stability in pasteurized milk in polyethylene-coated cartons and in polyethylene bottles. *Journal of the Society of Dairy Technology* **38**, 48–52.

Skibsted, L. H. (2000). Light induces changes in dairy products. *Bulletin of the International Dairy Federation* **345**, 4–9.

Thomas, E. L. (1981). Trends in milk flavours. *Journal of Dairy Science* **64**, 1023–1027.

Toyosaki, T., Yamamoto, A., and Mineshita, T. (1988). Kinetics of photolysis of milk riboflavin. *Milchwissenschaft* **43**, 143–146.

Valero, E., Villamiel, M., Sanz, J., and Martinez-Castro, I. (2000). Chemical and sensorial changes in milk quality on keeping the quality of pasteurized milk. *Letters in Applied Microbiology* **20**, 164–167.

Van Aardt, M., Duncan, J. S. E., Marcy, E., Long, T. E., and Hackey, C. R. (2001). Effectiveness of poly(ethylene terephthalate) and high density polyethylene in protection pf milk flavor. *Journal of Dairy Science* **84**, 1341–1347.

Vassila, E., Badeka, A., Kondyli, E., Savvaidis, I., and Kontominas, M. G. (2002). Chemical and microbiological changes in fluid milk as affected by packaging conditions. *International Dairy Journal* **12**, 715–722.

Zahar, M. and Smith, D. E. (1990). Vitamin A quantification in fluid dairy products: Rapid method for Vitamin A extraction for high performance liquid chromatography. *Journal of Dairy Science* **73**, 3402–3407.

Zygoura, P., Moyssiadi, T., Badeka, A., Kondyli, E., Savvaidis, I., and Kontominas, M. G. (2004). Shelf life of whole pasteurized milk in Greece: Effect of packaging material. *Food Chemistry* **87**, 1–9.

3

Aguilera, M. P., Beltran, G., Ortega, D., Fernandez, A., Jimenez, A., and Uceda, M. (2005). Characterization of virgin olive oil of Italian olive cultivars: Frantoio and Leccino, grown in Andalusia. *Food Chemistry* **89**(3), 387–391.

Araghipour, N., Colineau, J., Koot, A., Akkermans, W., Rojas, J. M. M., Beauchamp J., Wisthaler, A., Mark, T. D., Downey, G., Guillou, C., Mannina, L., and Ruth, S. (2008). Geographical origin classification of olive oils by PTR-MS. *Food Chemistry* **108**(1), 374–383.

Aranda, F., Gómez-Alonso, S., Rivera del Álamo, R. M., Salvador, M. D., and Fregapane, G. (2004). Triglyceride, total and 2-position fatty acid composition of Cornicabra virgin olive oil: Comparison with other Spanish cultivars. *Food Chemistry* **86**(4), 485–492.

Benincasa, C., Lewis, J., Perri, E., Sindona, G., and Tagarelli, A. (2007). Determination of trace element in Italian virgin olive oils and their characterization according to geographical origin by statistical analysis. *Analytica Chimica Acta* **585**(2), 366–370.

Bianchi, G., Angerosa, F., Camera, L., Reiniero, F., and Anglani, C. (1993). Stable carbon isotope ratios (13C/12C) of olive oil components. *Journal of Agricultural and Food Chemistry* **41**(11), 1936–1940.

Boggia, R., Zunin, P., Lanteri, S., Rossi, N., and Evangelisti, F. (2002). Classification and class-modeling of "Riviera Ligure" extra-virgin olive oil using chemical-physical parameters. *Journal of Agricultural and Food Chemistry* **50**(8), 2444–2449.

Boschelle, O., Rogic, A., Kocjancic, D., and Conte,L. S. (1994). Composition of lipid fraction of two olive cultivar from the island of Cres (Croatia), observed at different ripening eves. *Rivista Italiana delle Sostanze Grasse* **LXXI**, 341–346.

Bronzini de Caraffa, V., Gambotti, C., Giannettini, J., Maury, J., Berti, L., and Gandemer, G. (2008). Using lipid profiles and genotypes for the characterization of Corsican olive oils. European *Journal of Lipid Science and Technology* **110**(1), 40–47.

Bucci, R., Magrí, D. A., Magrí, L. A., Marini, D., and Marini, F. (2002). Chemical authentication of extra virgin olive oil varieties by supervised chemometric procedures. *Journal of Agricultural and Food Chemistry* **50**(3), 413–418.

Cimato, A., Dello Monaco, D., Distante, C., Epifani, M., Siciliano, P., Taurino, A. M., Zuppa, M., and Sani, G. (2006). Analysis of single-cultivar extra virgin olive oils by means of an Electronic Nose and HS-SPME/GC/MS methods. *Sensors and Actuators B: Chemical* **114**(2), 674–680.

Cosio, M. S., Ballabio, D., Benedetti, S., and Gigliotti, C. (2006). Geographical origin and authentication of extra virgin olive oils by an electronic nose in combination with artificial neural networks. *Analytica Chimica Acta* **567**(2), 202–210.

D'Imperio, M., Mannina, L., Capitani, D., Bidet, O., Rossi, E., Bucarelli, F., Quaglia, G., and Segre, A. (2007). NMR and statistical study of olive oils from Lazio: A geographical, ecological and agronomic characterization. *Food Chemistry* **105**(3), 1256–1267.

Dhifi, W., Angerosa, F., Serraiocco, A., Oumar, I., Hamrouni, I., and Marzouk, B. (2005). Virgin olive oil aroma: Characterization of some Tunisian cultivars. *Food Chemistry* **93**(4), 697–701.

Di Bella, G., Maisano, R., La Pera, L., Lo Turco, V., Salvo, F., and Dugo, G. (2007). Statistical characterization of Sicilian olive oils from the Peloritana and Maghrebian zones according to the fatty acid profile. *Journal of Agricultural and Food Chemistry* **55**(16), 6568–6574.

Diaz, T. G., Durán Merás, I., Sánchez Casas, J., and Alexandre Franco, M. F. (2005). Characterization of virgin olive oils according to its triglycerides and sterols composition by chemometric methods. *Food Control* **16**(4), 339–347.

Dupuy, N., Le Dréau, Y., Ollivier, D., Artaud, J., Pinatel, C., and Kister, J. (2005). Origin of French virgin olive oil registered designation of origins predicted by chemometric analysis of synchronous excitation-emission fluorescence spectra. *Journal of Agricultural and Food Chemistry* **53**(24), 9361–9368.

E C Regulation 2568/91 (1991). Characteristics of olive oil and olive-residue oil and on the relevant methods of analysis. *Official Journal of the European Communities* **L248**, 1–83.

E C Regulation 1187/2000 (2000). Register of protected designations of origin and Protected geographical indications. *Official Journal of the European Communities* **L133**, 19–20.

E C Regulation 1989/2003 (2003). Characteristics of olive and olive-pomace oil and on the relevant methods of analysis. *Official Journal of the European Communities* **L295**, 57–66.

Ferreiro, L. and Aparicio, R. (1992). Influence of altitude in the chemical composition of virgin olive oils from Andalusia. *Grasas y Aceites* **43**(3), 149–156.

Fontanazza, G., Patumi, M., Solinas, M., and Serraiocco, A. (1994). Influence of cultivars on the composition and quality of olive oil. *Acta Horticulturae* **356**, 358–361.

Galtier, O., Dupuy, N., Le Dréau, Y., Ollivier, D., Pinatel, C., Kister, J., and Artaud, J. (2007). Geographic origins and compositions of virgin olive oils determinated by chemometric analysis of NIR spectra. *Analytica Chimica Acta* **595**(1–2), 136–144.

Gutierrez, F., Arnaud, T., and Albi, M. A. (1999). Influence of ecological cultivation on virgin olive oil quality. *Journal of the American Oil Chemists' Society* **76**(5), 617–621.

Haddada, F. M., Manaï, H., Oueslati, I., Daoud, D. Sánchez, J., Osorio, E., and Zarrouk, M. (2007). Fatty acid, triacylglycerol and phytosterol composition in six Tunisian olive varieties. *Journal of Agricultural and Food Chemistry* **55**(26), 10941–10946.

Kiritsakis, A., Kanavouras, K., and Kiritsakis, K. (2002). Chemical analysis, quality control and packaging issues of olive oil. *European Journal of Lipid Science and Technology* **104**(9–10), 628–638.

Lorenzo, I. M., Pavon, J. L. P., Laespada, M. E. F., Pinto, C. G., and Cordero, B. M. (2002). Detection of adulterants in olive oil by headspace-mass spectrometry. *Journal of Chromatography A* **945**(1–2), 221–230.

Los, D. A. and Murata, N. (1998). Structure and expression of fatty acid desaturases. *Biochimica et Biophysica Acta* **1394**(1), 3–15.

Luna, G., Morales, M. T., and Aparicio, R. (2006). Characterization of 39 varietal virgin olive oils by their volatile compositions. *Food Chemistry* **98**(2), 243–252.

Mignani, A. G., Ciaccheri, L., Cimato, A., Attilio, C., and Smith, P. R. (2005). Spectral nephelometry for the geographic classification of Italian extra virgin olive oils. *Sensors and Actuators B: Chemical* **111–112**, 363–369.

Minguez-Mosquera, M. I., Rejano, L., Candul, B., Sanchez, A. H. and Garrido, J. (1991). Color-pigment correlation in virgin olive oil. *Journal of the American Oil Chemists' Society* **68**(5), 332–336.

Ollivier, D., Artaud, J., Pinatel, C., Durbec, J. P., and Guérère, M. (2003). Triacylglycerol and fatty acid compositions of French virgin olive oils. Characterization by chemometrics. *Journal of Agricultural and Food Chemistry* **51**(19), 5723–5731.

Oueslati, I., Haddada, F. M., Manai, H., Zarrouk, W., Taamalli, W., Fernandez, X., Lizzani-Cuvelier, L., and Zarrouk, M. (2008). Characterization of volatiles in virgin olive oil produced in the Tunisian area of Tataouine. *Journal of Agricultural and Food Chemistry* **56**(17), 7992–7998.

Owen, R. W., Haubner, R., Wurtele, G., Hull, W. E., Spiegelhalder, B. and Bartsch, H. (2004). Olives and olive oil in cancer prevention. *European Journal of Cancer Prevention* **13**(4), 319–326.

Petrakis, P. V., Agiomyrgianaki, A., Christophoridou, S., Spyros, A., and Dais, P. (2008). Geographical characterization of Greek virgin olive oils (cv. Koroneiki) using 1H and 31P NMR fingerprinting with canonical discriminant analysis and classification binary trees. *Journal of Agricultural and Food Chemistry* **56**(9), 3200–3207.

Stefanoudaki, E., Kotsifaki, F., and Koutsaftakis, A. (1997). The potential of HPLC triglyceride profiles for the classification of Cretan olive oils. *Food Chemistry* **60**(3), 425–432.

Stefanoudaki, E., Kotsifaki, F., and Koutsaftakis, A. (1999). Classification of virgin olive oils of the two major Cretan cultivars based on their fatty acid composition. *Journal of the American Oil Chemists' Society* **76**(5), 623–626.

Temime, S. B., Manai,H., Methenni, K., Baccouri, B., Abaza, L., Daoud, D., Casas, J. S., Bueno, E. O., and Zarrouk, M. (2008). Sterolic composition of Chétoui virgin olive oil: Influence of geographical origin. *Food Chemistry* **110**(2), 368–374.

Tsimidou, M. and Karakostas, K. X. (1993). Geographical classification of Greek virgin olive oil by non-parametric multivariate evaluation of fatty acid composition. *Journal of the Science of Food and Agriculture* **62**(3), 253–257.

Tsimidou, M., Macrae, R., and Wilson, I. (1987). Authentication of virgin olive oils using principal component analysis of triglyceride and fatty acid profiles: Part 1 – Classification of Greek olive oils. *Food Chemistry* **25**(3), 227–239.

Tura, D., Failla, O., Bassi, D., Redo, S., and Serraiocco, A. (2008). Cultivar influence on virgin olive (*Olea Europea* L.) oil flavor based on aromatic compounds and sensorial profile. *Scientia Horticulturae* **118**(2), 139–148.

4

Andrews, L. S., Marshall, D. L., and Grodner, R. M. (1995). Radiosensitivity of *Listeria monocytogenes* at various temperatures and cell concentrations. *Journal of Food Protection* **58**, 748–751.

AOAC (1995). Official methods of analysis (16th ed.), Arlington: Association of Official Analytical Chemists.

Badeka, A. B., Pappa, K., and Kontominas, M.G. (1999). Effect of microwave versus conventional heating on the migration of dioctyl adipate and acetyl tributyl citrate plasticizers from food grade PVC and P(VDC/VC) films into fatty foodstuffs. *Zeitschrift fur Lebensmittel-Untersuchung und Forschung A* **208**, 429–433.

Brody, A. L. and Marsh, K. S. (1997). *The Wiley Encyclopedia of Packaging Technology*. (2nd edit.). John Wiley and Sons, New York.

Buchalla, R., Boess, C., and Bogl, K. W. (2000). Analysis of volatile radiolysis products in gamma--irradiated LDPE and polypropylene films by thermal desorption--gas chromatography--mass spectrometry. *Applied Radiation and Isotopes* **52**, 251–269.

Buchalla, R., Schüttler, C., and Bogl, K. W. (1993a). Effects of ionizing radiation on plastic food packaging materials: A Review. Part 1: Chemical and Physical Changes. *Journal of Food Protection* **56** (11), 991–997.

Buchalla, R., Schüttler, C., and Bogl, K. W. (1993b). Effects of ionizing radiation on plastic food packaging materials: A Review. Part 2: Global migration, sensory changes and the fate of additives. *Journal of Food Protection* **56**(11), 998–1005.

Cano, J. M., Marin, M. L., Sanchez, A., and Hernandis, V. (2002). Determination of adipate plasticizers in poly (vinyl chloride) by microwave- assisted extraction. *Journal of Chromatography A* **963**, 401–409.

Castle, L., Gilbert, J., Jickells, S. M., and Gramshaw, J. W. (1988a). Analysis of the plasticizer acetyl tributyl citrate in foods by stable isotope dilution gas chromatography-mass spectrometry. *Journal of Chromatography* **437**, 281–286.

Castle, L., Jickells, S. M., Sharman, M., Gramshaw, J. W., and Gilbert, J. (1988b). Migration of the plasticizer acetyl tributyl citrate from plastic film into foods during microwave cooking and other domestic use. *Journal of Food Protection* **51**(12), 916–919.

Cook, D. (2001). Presentation at Gulf and South Atlantic Fisheries Conference, Biloxi, MS.

Deschenes, L., Arbour, A., Brunet, F., Court, M. A., Doyon, G. J., Fortin, J., and Rodrique, N. (1995). Irradiation of a barrier film: Analysis of some transfer aspects. *Radiation Physics and Chemistry* **46**, 805–808.

E C, Health and Consumers Directorate (2009). Substances listed in EU Directives on plastics in contact with food. Available at: ec.europa.eu/food/

EEC (1990). Commission Directive 90/128/EEC of 23 February 1990 relating to plastic materials and articles intended to come into contact with foodstuffs. *Official Journal of the European Communities*. No L 75, 19–39.

FICDB (2010). Food irradiation clearances database maintained by the Food and Environmental Subprogramme of the Joint FAO/IAEA Division of Nuclear Techniques in Food and Agriculture. Available at: nucleus.iaea.org

Gheysari, Dj., Behjat, A., and Haji-Saeid, M. (2001). The effect of high-energy electron beam on mechanical and thermal properties of LDPE and HDPE. *European Polymer Journal* **37**, 295–302.

Goulas, A. E., Anifantaki, K. I., Kolioulis, D. G., and Kontominas, M. G. (2000). Migration of di-(2-ethylhexyl) adipate plasticizer from food-grade PVC film into hard and soft cheeses. *Journal of Dairy Science* **83**, 1712–1718.

Goulas, A. E. and Kontominas, M. G. (1996). Migration of dioctyladipate plasticizer from food-grade PVC film into chicken meat products: effect of γ- radiation. *Zeitschrift fur Lebensmittel- Untersuchung und Forschung* **202**, 250–255.

Goulas, A. E., Riganakos, K. A., and Kontominas, M. G. (2004). Effect of electron beam and gamma radiation on the migration of plasticizers from flexible food packaging materials into foods and food simulants. In: V. Komolprasert, and K. M. Morehouse (Eds.), *Irradiation of food and packaging-Recent developments*. ACS Symposium Series No 875, Washington, DC. pp. 290–304.

Goulas, A. E., Zygoura, P., Karatapanis, A., Georgantelis, D., and Kontominas, M. G. (2007). Migration of di (2-ethylhexyl) adipate and acetyl tributyl citrate plasticizers from food-grade PVC film into sweetened sesame paste: Kinetic and penetration study. *Food and Chemical Toxicology* **45**, 585–591.

Grob, K., Pfenninger, S., Pohl, W., Laso, M., Imhof, D., and Rieger, K. (2007). European legal limits for migration from food packaging materials: 1. Food should prevail over simulants 2. More realistic conversion from concentrations to limits per surface area. PVC cling films in contact with cheese as an example. *Food Control* **18**, 201–210.

Heath, J. L. and Reilly, M. (1981). Migration of acetyl-tributylcitrate from plastic film into poultry products during microwave cooking. *Poultry Science* **60**(10), 2258–2264.

Komolprasert, V. and Morehouse, K. M. (2004). Irradiation of Food and Packaging-Recent Developments. American Chemical Society, Washington.

Kondyli, E., Demertzis, P. G., and Kontominas, M. G. (1992). Migration of dioctylphthalate and dioctyladipate plasticizers from food-grade PVC films into ground-meat products. *Food Chemistry* **45**, 163–168.

Lau, O. W., and Wong, S. K. (1996). The migration of plasticizers from cling film into food during microwave heating-effect of fat content and contact time. *Packaging Technology and Science* **9**, 19–27.

Loaharanu, P. (2003). Irradiated Foods. (5th ed.). American Council on Science and Health, New York.

MAFF (1991): Ministry of Agriculture, Fisheries and Food. Plasticizers: Continuing Surveillance. *Food Surveillance Paper No. 30*, Her Majesty's Stationery Office, London, UK.

Mercer, A., Castle, L., Comyn, J., and Gilbert, J. (1990). Evaluation of a predictive mathematical model of di-(2-ethylhexyl) adipate plasticizer migration from PVC film into foods. *Food Additives and Contaminants*, **7**(4), 497–507.

Molins, R. A. (2001). *Food Irradiation: Principles and Applications*. John Wiley and Sons, New York.

Murano, E.A. (1995). *Food Irradiation-A Sourcebook*. Iowa State University Press, US.

Nerin,C., Gancedo, P., and Cacho, J. (1992). Determination of bis(2-ethylhexyl) Adipate in Food Products. *Journal of Agricultural and Food Chemistry* **40**, 1833–1835.

Palumbo, S., Jenkins, R. K., Buchanan, R. L., and Thayer, D. W. (1986). Determination of irradiation D-values for Aeromonas Hydrophila. *Journal of Food Protection* **49**, 189–191.

Petersen, J. H. and Breindahl, T. (2000). Plasticizers in total diet samples, baby food and infant formulae. *Food Additives and Contaminants* **17**, 133–141.

Petersen, J. H., Naamansen, E. T., and Nielsen, P. A. (1995). PVC cling film in contact with cheese: health aspects related to global migration and specific migration of DEHA. *Food Additives and Contaminants* **12**(2), 245–253.

Poole, S. E., Mitchell, G. E., and Mayze, J. L. (1994). Low dose irradiation affects microbiological and sensory quality of sub- tropical seafood. *Journal of Food Science* **59**(1), 85–87, 105.

Sears, J. K. and Darby, J. R. (1982). *The technology of plasticizers*. John Wiley and Sons, Canada.

Sendon-Garcia, R., Sanches-Silva, A., Cooper, I., Franz, R., and Paseiro-Losada, P. (2006). Revision of analytical strategies to evaluate different migrants from food packaging materials. *Trends in Food Science and Technology* **17**, 1–13.

Startin, J. R., Sharman, M., Rose, M. D., Parker, I., Mercer, A. J., Castle, L., and Gilbert, J. (1987). Migration from plasticized films into foods. Migration of di-(2-ethylhexyl) adipate from PVC films during home use and microwave cooking. *Food Additives and Contaminants* **4**(4), 385–398.

Thibault, C. and Charbonneau, R. (1991). Shelf-life extension of fillets of Atlantic cod (*Gadus morhua*) by treatment with ionizing rays. *Sciences des Aliments* **11**(2), 249–261.

Till, D. E., Reid, R. C., Schwartz, P. S., Sidman, K. R., Valentine, J. R., and Whelan, R. H. (1982). Plasticizer migration from polyvinyl chloride film to solvents and foods. *Food and Chemical Toxicology* **20**, 95–104.

Van Lierop, J. B. H. and Van Veen, R. M. (1988). Determination of plasticizers in fat by gas chromatography-mass spectrometry. *Journal of Chromatography* **447**, 230–233.

Wei, D. Y., Wang, M. L., Guo, Z. Y., Wang, S., Li, H. L., Zhang, H. N., and Gai, P. P. (2009). GC/MS method for the determination of adipate plasticizers in ham sausage and its application to kinetic and penetration studies. *Journal of Food Science* **74**(5), 392–398.

Wilkinson, V. M. and Gould, G. W. (1996). *Food Irradiation-A Reference Guide*. Butterworth–Heinemann, Oxford.

Zygoura, P. D., Goulas, A. E., Riganakos, K. A., and Kontominas, M. G. (2007). Migration of di-(2-ethylhexyl) adipate and acetyl tributyl citrate plasticizers from food-grade PVC film into isooctane: Effect of gamma radiation. *Journal of Food Engineering* **78**, 870–877.

5

Amerine, M. A. and Ough, C. S. (1974). *Wine and Must Analysis*. John Wiley & Sons, Inc, New York, USA.

Arora, D. K., Hansen, A. P., and Armagost, M. S. (1991). Sorption of flavor compounds by low density polyethylene film. *Journal of Food Science* **56**, 1421–1423.

Ayhan, Z., Yeom, H. W., Zhang, Q. H., and Min, D. B. (2001). Flavor, color and vitamin C retention of pulsed electric field processed orange juice in different packaging materials. *Journal of Agricultural and Food Chemistry* **49**, 669–674.

Bach, H. P. and Hess, K. H. (1984). Einflus der Verpackungsart auf der Wein. *Weinwirtsch Technology* **5**, 121.

Bayouone, C., Baumes, R., Crouzet, J., and Gunata, Z. (2000). Aromas Los constituyentes aromaticos de la fase prefermentativa. In *Enologia. Fundamentos Cientificos Y Tecnologicos*. AMV Ediciones, Madrid, pp.147–158.

Boulton, R. B., Singleton, V. L., Bisson, L. F., and Kunkee, R. E. (1996). *Principles and Plastics of Winemaking*. Chapman & Hall, New York, USA.

Cantagrel, P., Mazerolles, G., Vidal, J. P., Galy, B., Boulesteix, J. M., Lablanquie, O., and Gaschet, J. (1998). *Analytical and Organoleptic*

Changes in Cognac During Ageing. Part 1. Rivista Italiana EPPOS, pp. 406–424.

Cantagrel, R., Lurton, L., Vidal, J. P., and Galy, B. (1997). Form wine to Cognac. In *Fermented Beverage Production.* A. G. H. Lea and Piggott J. R. (Eds.), Blackie Academic and Professional, London, UK.

Charara, Z. N., Williams, J. W., Scmidt, R. H., and Marshall, M. R. (1992). Orange flavor absorption into various polymeric packaging materials. *Journal of Food Science* **57**, 963–966.

Cole, V. C. and Noble, A. C. (1997). Flavor chemistry and assessment. In *Fermented Beverage Production.* A. G. H. Lea and J. R. Piggott (Eds.). Blackie Academic and Professional, London, UK, pp. 361–385.

Diez Marques, C., Coll Hellin, L., Gutiereze Ruiz, L., and Zapata Revilla, A. (1994). Analytical study of apple liquers. *Zeitschrift Fur Lebensmittel-Untersuschung Und-Forschung* **198**, 60–65.

Falque, E., Fernandez, E., and Dubourdieu, D. (2001). Differentiation of white wines by their aromatic index. *Talanta* **54**, 271–281.

Ferreira, V., Escudero, A., Fernandez, P., and Cacho, J. F. (1997). Changes in the profile of volatile compounds in wines stored under oxygen and their relationshiphs with the vrowing process. *Zeitschrift Fur Lebensmittel-Untersuchung Und-Forschung A*, **205**, 392–396.

Ferreira, V., Fernandez, P., Pena, C., Escudero, A., and Chacho, J. F. (1995). Investigation on the role played by fermentation esters in the aroma of young Spanish wines by multivariate analysis. *Journal of the Science of Food and Agriculture* **67**, 381–392.

EEC (1990). Commission Regulation (EEC) No. 2676/90 determining Community methods for the analysis of wines. *Official Journal* **L272**, 1–192.

Godden, P., Francis, L., Field, J, Gishen, M., Coulter, A., Valente, P., Hoj, P., and Robinson, E. (2001). Wine bottle closures: Physical characteristics and effect on composition and sensory properties of a Semillon wine. Performance up to 20 months post-bottling. *Australian Journal of Grape and Wine Research* **7**, 64–105.

Halek, G. W. and Luttman, J. P. (1991). Sorption bahavior of citrus-flavor compounds in polyethylenes and polyprolpylenes: Effects of permeant functional groups and polymer structure. In *Food and Packaging Interactions II.* S. L. Risch and J. H. Hotchkiss (Eds.). ACS Symposium Series 473, American Chemical Society, Washington, USA, pp. 212–226.

Ikegami, T., Nagashima, K., Shimoda, M., Tanaka, Y., and Osajima, Y. (1991). Sorption of volatile compounds in aqeous solution by ethylene-vinyl alcohol copolymer films. *Journal of Food Science* **56**, 500–503, 509.

Jackson, R. S. (2000). *Wine Science: Principles, Practice, Perception.* Academic Press, 2nd ed. Santiago, Chapter 6, pp. 232–280.

Kana, K., Kanellaki, M., Kouinis, J., and Koutinas, A. A. (1988). Alcohol production from raisin extracts: Volatile by-products. *Journal of Food Science* **53**, 1723–1724, 1749.

Kontominas, M. G. (2010). Effects of packaging in milk quality and safety. In *Improving the safety and quality of milk.* M. W. Griffiths (Ed.). Woodhead publishing limited and CRC press, Cambridge, UK, pp. 136–155.

Li, H., Guo, A., and Wang, H. (2008). Mechanisms of oxidative browning of wine. *Food Chemistry* **108**(1), 1–13.

Linssen, J. P. H. and Roozen, J. P. (1994). Food Flavor and Packaging Interactions. In *Food Packaging and Preservation.* M. Mathlouthi (Ed.). Blackie Academic & Professional, Glasgow, UK, pp. 48–60.

Linssen, J. P. H., Verheul, A., Roozen, J. P., and Posthummus, M. A. (1991). Absorption of flavor compounds by packaging material: drink yogurts in polyethylene bottles. *International Dairy Journal* **1**, 33–40.

Loyaux, D., Roger, S., and Adda, J. (1981). The evolution of champagne volatiles during ageing. *Journal of the Science and Food Agriculture* **32**, 1254–1258.

Mahoney, S. M., Hernandez, R. J., Giacin, J. R., Harte, B. R., and Miltz, J. (1988). Permeability and solubility of d-limonene vapor in cereal package liners. *Journal of Food Science* **53**, 253–257.

Maicas, S., Gil, J. V., Pardo, I., and Ferrer, S. (1999). Improvement of volatile composition of wines by controlled addition of malolactic bacteria. *Food Research International* **32**, 491–496.

Mangas, J., Rodriquez, R., Moreno, J., and Blanco, D. (1996a). Volatiles in distillates of cider aged in American oak wood. *Journal of Agricultural and Food Chemistry* **44**, 268–273.

Mangas, J., Rodriquez, R., Moreno, J., and Blanco, D. (1996b). Changes in the major volatile compounds of cider distillates during maturation. *Lebensmittel-Wissenchaft Und-Technologie* **29**, 357–364.

Meilgaard, M. C. (1981). Beer Flavor. *Dissertation.* Technical University of Denmark. University of Michigan, Ann Arbor. Michigan, University Microfilms International, USA.

Mentana, A., Pati, S., La Notte, E., and Del Nobile, M. A. (2009). Chemical changes in Apulia table wines as affected by plastic packages. *Food Science and Technology* **42**, 1360–1366.

Monica-Lee, K., Paterson, A., and Piggott, J. R. (2000). Perception of whisky flavor reference compounds by Scottish distillers. *Journal of the Institute of Brewing* **106**, 203–208.

Moshanos, M. G. and Shaw, P. E. (1998). Changes in composition of volatile constituents in aseptically packaged orange juice. *Journal of Agricultural and Food Chemistry* **37**, 157–161.

Muratore, G., Lanza, C. M., Baiano, A., Tamagnone, P., Ausmundo, C., and Del Nobile, M. A. (2005). The influence of using different packaging on the quality decay Kinetics of Cuccia. *Journal of Food Engineering* **73**, 239–245.

Nielsen, T. J. and Jägerstand, I. M. (1994). Flavor scalping by food packaging. *Trends in Food Science and Technology* **5**, 353–356.

Nielsen, T. J., Jägerstand, I. M., and Oste, R. E. (1992). Study of factors affecting the absorption of aroma compounds into low-density polyethylene. *Journal of the Science and Food Agriculture* **60**, 377–381.

Perez-Coello, M. S., Gonzalez-Vinas, M. A., Garcia-Romero, E., Diaz-Maroto, M. C., and Cabeluzo, M. D. (2003). Influence of storage temperature on the volatile compounds of young white wine. *Food Control* **14**, 301–306.

Perez-Prieto, L. J., Lopez-Roca, J. M., and Gomez-Plaza, E. (2003) Differences in major volatile compounds of red wines according to storage length and storage conditions. *Journal of Food Composition and Analysis* **16**, 697–705.

Pieper, G., Borgudd, L., Ackermann, P., and Fellers, P. (1992). Absorption of aroma volatiles of orange juice into laminated carton packages did not affect sensory quality. *Journal of Food Science* **57**, 1408–1411.

Rapp, A. and Mandery, H. (1986). Wine aroma. *Experientia* **42**, 873–884.

Robertson, G. L., (2006). *Food Packaging. Principles and Practise*. 2nd ed., Taylor and Francis Group, Boca Raton, USA.

Ribereau-Gayon, P. and Peynand, E. (1966). Traite d' Oenologie II. *Composition, Tranformation et Traiments des Vins*. Dunod Paris, pp. 1065.

Salame, M. (1989). The use of barrier polymers in food and beverage packaging. In *Plastic film technology, vol. 1, high barrier plastic films for packaging*. K. M. Finlayson (Ed.). Technologic Publishing Co. Inc., Lancaster, USA, pp. 132–145.

Shimoda, M., Ikegami, T., and Osajima, Y. (1988). Sorption of flavor compounds in aqueous solution on adhesive and cohesive bond strengths in laminations. *Journal of Plastic Film and Sheets* **6**(3), 232–246.

Shinohara, M. and Shimuzu, J. (1981). *Nippon Nogeikagaku Kaishi*. **55**, 679–685.

Silva, M. L. and Malcata, F. X. (1998). Relationships between storage conditions of grape pomace and volatile composition of spirits obtained there from. *American Journal of Enology and Viticulture* **59**, 56–64.

Silva, M. L., Macedo, A. C., and Malcata, F. X. (2000). Review: Steam distilled spirits from fermented grape pomace. *Food Science and Technology International* **6**, 285–300.

Skouroumounis, G. K., Kwiatkowski, M., Sefton, M. A., Gawel, R., and Waters, E. J. (2003). In situ measurement of white wine absorbance in clear and in colored bottles using a modified laboratory spectrophotometer. *Australian Journal of Grape and Wine Research* **9**(2), 138–148.

Sobek, A. (2003). Evaluation of storage stability of multilayer PET bottles. *Fluessiges Obs* **70**(12), 708–711.

Tawfik, M. S., Devlieghere, F., and Huyghebaert, A. (1998). Influence of d-limonene absorption on the physical properties of refillable PET. *Food Chemistry* **61**(1), 157–162.

Van der Merwe, C. A. and Van Wyk, C. K. (1981). The contribution of some fermentation products to the odor of dry white wines. *American Journal of Enology and Viticulture* **32**, 41–46.

Versini, G. (1993). Volatile compounds of spirits. In B. Dineche (Ed.), *Les Acquisitions Recentes en Chromatographie du Vin Cours Europeen de Formation Continue*, 31 Mars 1, 2 et 3 Avril 1992 Paris: Tec & Dic-Lavoisier, Porto, Portugal. pp. 189–213.

Willige van, R. W., Schoolmeester, D., Ooij, A. V., Linssen, J., and Voragen, A. (2002). Influence of storage time and temperature on absorption of flavor compounds from solutions by plastic packaging materials. *Jourmal Food Science* **67**(6), 2023–2031.

6

Achen, M. (2000). Efficacy of ozone in inactivating *Escherichia coli* O157:H7 in pure cell suspensions and on apples. M.S. thesis. The Ohio State University, Columbus.

Aguayo, E., Escalona, V. H., and Artés, F. (2006). Effect of cyclic exposure to ozone gas on physicochemical, sensorial and microbial quality of whole and sliced tomatoes. *Postharvest Biology and Technology* **39**, 169–177.

Alonso, A., Vázquez-Araújo, L., García-Martínez, S., Ruiz, J. J., and Carbonell-Barrachina, A. A. (2009). Volatile compounds of traditional and virus-resistant breeding lines of Muchamiel tomatoes. *European Food Research and Technology* **230**, 315–323.

Anonymous (2004). Chemical Composition. In: *Fruit and vegetable processing*—Chapter 02 General properties of fruit and vegetables; chemical composition and nutritional aspects; structural features.

Baranovskaya, V. A., Zapolskii, O. B., Ovrutskaya, I. Y., Obodovskaya, N. N., Pschenichnaya, E. E., and Yushkevich, O. I. (1979). Use of ozone gas sterilization during storage of potatoes and vegetables. Konservn. *Ovoshchesus Promst* **4**, 10–12.

Baldwin, E. A., Nisperos-Carriedo, M. O., and Moshonas, M. G., (1991). Quantitative analysis of flavor and other volatiles and certain constituents of two tomato cultivars during ripening. *Journal of the American Society for Horticultural Science* **116**, 265–269

Ball, J. A. (1997). Evaluation of two lipid-based edible coatings for their ability to preserve post harvest quality of green bell peppers. M. Sc. Thesis, Faculty of the Virginia Polytechnic Institute and State University, Blacksburg, Virginia, p 89.

Barth, M. M., Zhou, C., Mercier, M., and Payne, F. A. (1995). Ozone storage effects on anthocyanin content and fungal growth in blackberries. *Journal of Food Science* **60**, 1286–1287.

Batu, A. and Thompson, A. K. (1998). Effects of modified atmosphere packaging on post harvest qualities of pink tomatoes. *Turkish Journal of Agriculture and Forestry* **22**, 365–372.

Bhattacharya, J., Goldman, D., and Sood, N. (*2004*). Price Regulation in Secondary Insurance Markets. *Journal of Risk and Insurance* **21**(4), 643–675.

Buttery, R. G., Seifert, R. M., Guadagni, D. G., and Ling, L. C., (1971). Characterization of additional volatile components of tomato. *Journal of Agricultural and Food Chemistry* **19**, 524–529.

Buttery, R. G., Teranishi, R., and Ling, L. C. (1987). Fresh tomato aroma volatiles: A quantitative study. *Journal of Agricultural and Food Chemistry* **35**, 540–544.

Buttery, R. G., Teranishi, R., Ling, L. C., Flath, R. A., and Stern, D. J., (1988). Quantitative studies on origins of fresh tomato aroma volatiles. *Journal of Agricultural and Food Chemistry* **36**, 1247–1250.

Buttery, R. G. and Ling, L. C. (1993). Volatile components of tomato fruit and plant parts: relationship and biogenesis. In *Bioactive Volatile Compounds from Plants*. R. Teranishi, R. G. Buttery, and H. Sugisawa (Eds.). ACS, Washington, pp. 22–33.

Chiesa, L., Diaz, L., Cascone, O., Pańak, K., Camperi, S., Frezza, D., and Fragaus, A. (1998). Texture changes on normal and long shelf life of tomato (*Lycopersicon esculentum* Mill.) fruit ripening. *Actas de Horticultura (ISHS)* **464**(1), 487–488.

Choi, L. H. and Nielsen, S. S. (2005). The effects of thermal and non thermal processing methods on apple cider quality and consumer acceptability. *Journal of Food Quality* **28**, 13–29.

Coke, A. L. (1993). *Mother nature's best remedy: Ozone, Water Conditioning and Purification*, (October), 48–51.

Cole, E. R. and Kapur, N. S. (1957). The stability of lycopene. I. Degradation by oxygen. II. Oxidation during heating of tomato pulps. *Journal of the Science of Food and Agriculture* **8**, 360–366.

Dalal, K. B., Olson, L. E., Yu, M. H., and Salunkhe, D. K. (1967). Gas chromatography of the field, glasshouse-grown and artificially ripened tomatoes. *Phytochemistry* **95**, 1–55.

Dosti, B. (1998). Effectiveness of ozone, heat and chlorine for destroying common food spoilage bacteria in synthetic media and biofilms. Thesis. Clemson University, Clemson, SC, p. 69.

Eriksson, C. E. (1968). Alcohol NAD oxidoreductase from peas. *Journal of Food Science* **33**, 525.

Ewell, A. W. (1950). Ozone and its applications in food preservation. Refrigerating Engineering Application Data, section 50, published in Refrigerating Engineering, section 2, September.

Fetner, R. H. and Ingols, R. S. (1956). A comparison of the activity of ozone and chlorine against *Escherichia coli* at 1°C. *Journal of General Microbiology* **15**, 381–385

Foegeding, P. M. (1985). Ozone inactivation of *Bacillus* and *Clostridium* spore populations and the importance of the spore coat to resistance. *Food Microbiology* **2**, 123–134.

Gane, R. (1936). The respiration of bananas in presence of ethylene. *New Phytologist* **36**, 170–178.

Garcia, J. M., Castellano, J. M., Nadas, A., and Olias, J. M. (1998). Effect of ozone on citrus cold storage. *Cost* **1**, 237–241.

Gejima, Y., Zhang, H., and Nagata, M. (2003). Judgment on level of maturity for tomato quality using L* a* b* Color image processing. *Proc IEEE/ASME International Conference on advanced intelligent Melectronics AIM (2003)* **2**, 1355–1359.

Guzel-Seydim, Z. B. (1996). The use of ozonated water as a cleaning agent in dairy processing equipment. Thesis.

Guzel-Seydim, Z., Bever, P., and Greene, A. K. (2004a). Efficacy of ozone to reduce bacterial populations in the presence of food components. *Food Microbiology* **21** 475–479.

Guzel-Seydim, Z., Greene, A. K., and Seydim, A. C. (2004b). Use of ozone in the food industry. *Lebensmittel-Wissenschaft & Technologie* **37**, 453–460.

Heath, H. B. (*1978*). *Flavor technology*: Profiles, products, applications. Westport, CT: AVI Publishing Company, p. 5–42.

Ho, C. and Ichimura, N. (1982). Identification of Heterocyclic Compounds in the Volatile Flavour of Fresh Tomato. *Lebensmittel-Wissenschaft & Technologie* **15**, 340–342.

Ishizaki, K., Shinriki, N., and Matsuyama, H. (1986). Inactivation of *Bacillus* spores by gaseous ozone. *Journal of Applied Microbiology* **60**, 67–72.

Jin, L., Xiaoyu, W., Honglin, Y., Zonggan, Y., Jiaxun, W., and Yaguang, L., (1989). Influence of discharge products on post-harvest physiology of fruit. In *Proceedings of the Sixth International Symposium on High Voltage Engineering, 28 August to 1 September*, New Orleans, LA, p. 4.

Karaca, H. and Velioglu, Y. S. (2007). Ozone Applications in Fruit and Vegetable Processing. *Food Reviews International* **23**, 91–106.

Katayama, O., Tubata, K., and Yamato, I. (1967). The aroma components in fruits and vegetables. II. Volatile components of tomato. *Nippon Shokuhin Kogyo Gakkaishi* **14**, 4–44.

Kazeniac, S. J. and Hall, R. M. (1970). Flavor chemistry of tomato volatiles. *Journal of Food Science* **35**, 519–530.

Kim, M. J., Oh, Y. A., Kim, M. H., Kim, M. K., and Kim, S. D. (1993). Fermentation of Chinese cabbage kimchi inoculated with Lactobacillus acidophilus and containing ozone-treated ingredients. *Journal of Korean Society of Food and Nutrition* **22**, 165–174.

Kim, J. G., Yousef, A. E., and Dave, S., (1999). Application of ozone for enhancing the microbiological safety and quality of foods: A review. *Journal of Food Protection* **62**, 1071–1087.

Kogelschatz, U. (1988). Advanced ozone generation. In *Process technologies for water treatment.* S. Stucki (Ed.). Plenum Publishers, New York, pp. 87–120.

Koukol, J. and Corm, E. E. (1961). The metabolism of aromatic compounds in higher plants. IV. Purification and properties of the phenylalanine deaminase of Hordeum vulgare. *The Journal of Biological Chemistry* **236**, 26–92.

López Camelo, A. F. and Gómez, P. A. (2004). Comparison of color indexes for tomato ripening. *Horticultasileira* **22**, 534–537.

Maguire, Y. P. and Solver, G M. (1980). Influence of atmospheric oxygen and ozone on ripening indices of normal (Rin) and ripening inhibited (rin) tomato cultivars. *Journal of Food Biochemistry* **4**, 99–110.

Majchrowicz, A. (1998). Food safety technology: A potential role for ozone? Agricultural Outlook, Economic Research Service/USDA, pp. 13–15.

Marković K., Vahčic N., Kovačević Ganić K. and Banović M. (2007). Aroma volatiles of tomatoes and tomato products evaluated by solid-phase microextraction. *Flavour and Fragrance Journal* **22**, 395–400.

Mencarelli F. and Saltveit Jr. M.E. (1988). Ripening of mature-green tomato fruit slices. *Journal of the American Society for Horticultural Science* **113**(5), 742–745.

Meredith, E. and Purcell, A. E., (1966). Changes in the carotenes of ripening Homestead tomatoes. *Journal of the American Society for Horticultural Science* **89**, 544–548

Nadas, A., Olmo, M., and Garcia, J., M. (2003). Growth of *Botrytis cinerea* and Strawberry Quality in Ozone-enriched Atmospheres. *Journal of Food Science* **68**(5), 1798–1802.

Naitoh, S. and Shiga, I. (1989). Studies on utilizing of ozone in food preservation. IX. Effect of ozone treatment on elongation of hypocotyls and microbial counts of bean sprouts. *Journal of Japanese Society of Food Science and Technology* **36**, 181–188.

Nelson, P. E. and Hoff, J. E. (1969). Tomato Volatiles: Effect of Variety, Processing and Storage Time. *Journal of Food Science* **34**, 53–57.

Oehlschlaeger, H. F. (1978). Reactions of ozone with organic compounds. In *Ozone/chlorine dioxide oxidation products of organic material*. R. G. Rice and J. A. Cotruvo (Eds.). Ozone Press International, Cleveland, pp. 20–37.

Palou, L., Smilanick, J. L., Crisosto, C. H., and Mansour, M. (2001). Effect of gaseous ozone exposure on the development of green and blue molds on cold stored citrus fruit. *Plant Disease* **85**(6), 632–638.

Palou, L., Crisosto, C. H., Smilanick, J. L., Adaskaveg, J. E., and Zoffoli, J. P. (2002). effects of continuous 0.3 ppm ozone exposure on decay development and physiological responses of peaches and table grapes in cold storage. *Postharvest Biology and Technology* **24**, 39–48.

Palou, L., Crisosto, C. H., Smilanick, J. L., Adaskaveg, J. E., and Zoffoli, J. P. (2005). In *Evaluation of the Effect of Ozone Exposure on Decay Development and Fruit Physiological Behavior, Proceedings of the 4th International Conference on Postharvest. 26–31March, 2000.* R. Ben-Arie and S. Phiosoph-Hadas (Eds.). ISHS 553, in Central Valley Postharvest Newsletter, University of California, Acta Hort, **14**(1), pp. 1–10.

Parish, M. E., Beuchat, L. R., Suslow, T. W., Harris, L. J., Garret, E. H., Farber, J. N., and Busta, F. F. (2003). Methods to reduce eliminate pathogens from fresh and fresh-cut produce. *Comprehensive Reviews in Food Science and Food Safety* **2**, 161–173.

Perez, A. G., Sanz, C., Rios, A., Olias, J. J., and Olias, J. M. (1999). Effects of ozone treatment on postharvest strawberry quality. *Journal of Agricultural and Food Chemistry* **47**, 1652–1656.

Petró-Turza, M. (1987). Flavor of Tomato and Tomato Products. *Food Reviews International* **2**, 309–351.

Redd, J. B., Hendrix, C. M., and Hendrix, D. L. (1986). Quality control manual for citrus processing plants. In *Book 1*. James B. Redd, Donald L., Hendrix, and Charles Marion Hendrix (Eds.). Safety Harbor, FL, Intercity, Inc.

Restaino, L., Frampton, E. W., Hemphill, J. B., and Palnikar, P. (1995). Efficacy of ozonated water against various food-related microorganisms. *Applied and Environmental Microbiology* **61**(9), 3471–347.

Rice, R. G., Farquhar, J. W., and Bollyky, L. J. (1982). Review of the applications of ozone for increasing storage times of perishable foods. *Ozone Science and Engineering* **4**, 147–163.

Salvador, A., Abad, I., Arnal, L., and Martinez – Javega, J. M. (2006). Effect of ozone on postharvest quality of Persimmon. *Journal of Food Science* **71**, 443–446.

Sammi, S. and Masud, T. (2007). Effect of Different Packaging Systems on Storage Life and

Quality of Tomato (*Lycopersicon esculentum* var. Rio Grande) during Different Ripening Stages. *Journal of Food Safety* **9**, 37–44

Schermuller, J. and Kochmann, H. J. (1969). Analyses of volatile aromatic substances in tomatoes by gas chromatography and mass spectroscopy. *Z. Lebensmittel Untersuch Forsch* **141**, 1.

Servili, M., Selvaggini, R., Taticchi, A., Begliomini, A. L., and Montedoro, G. F. (2000). Relationships between the volatile compounds evaluated by solid phase microextraction and the thermal treatment of tomato juice: Optimization of the blanching parameters. *Food Chemistry* **71**, 407–415.

Shah, B. M., Salunkhe, D. K., and Olson, I. F. (1969). Effects of ripening process on chemistry of tomato volatiles. *Journal of the American Society for Horticultural Science* **94**, 171.

Shalluf, M. A., Tizaoui, C., and Karodia, N. (2007). Controlled atmosphere storage technique using ozone for delay ripening and extend the shelf life of tomato fruit. IOA Conference and Exhibition Valencia, October 29–31, Spain.

Sharma, S. K., Lemaguer, M., Liptay, A., and Pousa, V. (1996). Effect of composition on the rheological properties of tomato thin pulp. *Food Research International* **29**(2), 175–179.

Singh, N., Singh, R. K., Bhunia, A. K., and Stroshine, R. L. (2002). Efficacy of chlorine dioxide, ozone, and thyme essential oil or a sequential washing in killing *Escherichia coli* O157:H7 on lettuce and baby carrots. *Lebensmittel Wissenschaft and Technologie* **35**, 720–729.

Skog, L. J. and Chu, C. L. (2001). Effect of ozone on qualities of fruits and vegetables in cold storage. *Canadian Journal of Plant Science* **81**, 773–778.

Song, J., Fan, L., Forney, C. F., Jordan, M. A., Hildebrand, P. D., Kalt, W., and Ryan, D. A. J., (2003). Effect of ozone treatment and controlled atmosphere storage on quality and phytochemicals in high bush blueberries. In *Proceedings of the Eighth Controlled Atmosphere Research Conference, vol. II*. J. Oosterhaven and H. W. Peppelenbos (Eds.). Acta Horticultural, 600, ISHS, pp. 417–423.

Stevens, M. A. (1972). Citrate and malate concentrations in tomato fruits: Genetic control and maturational effects. *Journal of the American Society for Horticultural Science* **97**, 655–658.

Tandon, K. S. (1997). *Odor Thresholds and Flavor Quality of Fresh Tomatoes. Master Thesis.* The University of Georgia, Athens, GA.

Thomson, A. E., Tomes, M. L., and Wann, E. V. (1965). Characterization of crimson tomato fruit color. *Journal of the American Society for Horticultural Science* **86**, 610–616.

Tijskens, L. M. M. and Evelo, R. G. (1994). Modelling color of tomatoes during postharvest storage. *Postharvest Biology and Technology* **4**(1–2), 85–98.

Tiwari, B. K., Muthukumarappan, K., O'Donnell, C. P., and Cullen, P. J. (2008). Kinetics of freshly squeezed orange juice quality changes during ozone processing. *Journal of Agricultural & Food Chemistry* **56**, 6416–6422.

Tiwari, B. K., O' Donnell, C. P., Brunton, N. P., and Cullen, P. J. (2009). Degradation kinetics of tomato juice quality parameters by ozonation. International *Journal of Food Science and Technology* **44**, 1199–1205.,

Tzortzakis, N. G., Borland, A., Singleton, I., and Barnes, J. D. (2007). Impact of atmospheric ozone-enrichment on quality-related attributes of tomato fruit. *Postharvest Biology and Technology* **45**, 317–325.

US FDA (2001). United States Food and Drug Administration. Secondary direct food additives permitted in food for human consumption, final rule. *Federal Register* **66**, 33829–33830

Victorin, K. (1992). Review of the genotoxicity of ozone. *Mutation Research* **277**, 221–238.

Zhang, L., Lu, Z., Yu, Z., and Gao, X. (2005). Preservation Fresh-cut Celery by Treatment of Ozonated Water. *Food Control* **16**, 279–283.

7

Adam, K., Sivropoulou, A., Kokkini, S., Lanaras, T., and Arsenakis, M. (1998). Antifungal activities of *Origanum vulgare* subsp. *hirtum*, *Mentha spicata*, *Lavandula angustifolia* and *Salvia fru-*

ticosa essential oils against human pathogenic fungi. *Journal of Agricultural and Food Chemistry* **46**, 1739–1745.

Ahn, D. U., Olson, O. G., Jo, C., Chen, X., and Nu Cand Lee, J. I. (1998). Effect of muscle type, packaging and irradiation on lipid oxidation, volatile production and colour in raw pork patties. *Meat Science* **49**(1), 27–39.

Botsoglou, N. A., Grigoropoulou, S. M., Botsoglou, E., Govaris, A., and Papageorgiou, G. (2003). The effects of dietary oregano essential oil and a-tocopheryl acetate on lipid oxidation in raw and cooked turkey during refrigerated storage. *Meat Science* **65**, 1193–1200.

Burt, S. (2004). Essential oils: Their antibacterial properties and potential applications in foods-a review in press. *International Journal of Food Microbiology* **94**(3), 223–253.

Byun, J. S., Min, J. S., Kim, I. S., Kim, J. W., Chung, M. S., and Lee, M. (2003). Comparison of indicators of microbial quality of meat during aerobic cold storage. *Journal of Food Protection* **66**, 1733–1737.

Chouliara, I. and Kontominas, M. G. (2006). Combined effect of thyme essential oil and modified atmosphere packaging to extend shelf life of fresh chicken meat. In *Recent Progress in Medicinal Plants: Natural Product (15)*. J. N. Govil, V. K. Singh, K. Almad, and R. K. Sharma (Eds.). Studium Press, LLC, USA, pp. 423–442.

Chouliara, I., Savvaidis, I., Riganakos, K., and Kontominas, M. G. (2005). Shelf life extension of vacuum-packaged sea bream (*Sparus aurata*) fillets by combined γ-irradiation and refrigeration: Microbiological, chemical and sensory changes. *Journal of Science and Food Agriculture* **85**, 779–784.

Chouliara, E., Karatapanis, A., Savvaidis, I. N., and Kontominas, M. G. (2007). Combined effect of oregano essential oil and modified atmosphere packaging on shelf life extension of fresh chicken breast meat, stored at 4°C. *Food Microbiology* **24**, 607–617.

Cleveland, J., Montville, T. J., Nes, I. F., and Chikindas, M. L. (2001). Bacteriocins: Safe, natural antimicrobials for food preservation. *International Journal of Food Microbiology* **71**, 1–20.

Cutter, C. N. and Siragusa,G. R. (1996). Reductions of *Listeria innocua* and *Brochothrix thermosphacta* on beef following nisin spray treatments and vacuum packaging. *Food Microbiology* **13**, 23–33.

Dawson, P. L., Hon, H., Vollet, L. M., Clardy, L. B., Martinez, R. M., and Acton, J. C. (1995). Film oxygen transmission rate effects on ground chicken meat quality. *Poultry Science* **14**, 1381–1387.

Deans, S. G. and Richie, G. (1987). Antimicrobial properties of plant essential oils. *International Journal of Food Microbiology* **5**, 165–180.

Devlieghere, F., Vermeiren, L., and Debevere, J. (2004). New preservation technologies: Possibilities and limitations. *International Dairy Journal* **14**, 273–285.

Du, M., Hur, S. J., and Ahn, D. U. (2002). Raw meat packaging and storage affect the color and odor of irradiated breast fillets after cooking. *Meat Science* **61**, 49–54.

El-Alim, S. S. L., Lugasi, A., Havari, J., and Dworschek, E. (1999). Culinary herbs inhibit lipid oxidation in raw and cooked minced meat patties during storage. *Journal of Science and Food Agriculture* **79**, 277–285.

Fang, T. J. and Lin, L. W. (1994). Growth of *Listeria monocytogenes* and *Pseudomonas fragi* on cooked pork in a modified atmosphere packaging/nisin combination. *Journal of Food Protection* **57**, 479–485.

Farkas, J. (1990). Combination of irradiation with mild heat treatment, *Food Control* **1**, 223–229.

Giatrakou, V., Ntzimani, A., and Savvaidis, I. N. (2010). Effect of chitosan and thyme oil on a ready to cook chicken product. *Food Microbiology* **27**, 132–136.

Gill, A. O. and Holley, R. A. (2003). Interactive inhibition of meat spoilage and pathogenic bacteria by lysozyme, nisin and EDTA in the presence of nitrite and sodium chloride at 24°C. *International Journal of Food Microbiology* **80**(3), 251–259.

Goulas, A. E. and Kontominas, M. G. (2007). Combined effect of light salting, modified atmosphere packaging and oregano essential oil on the shelf life of sea bream (*Sparus aurata*):

Biochemical and sensory attributes. *Food Chemistry* **100**(1), 287–296.

ICMFS (1986). International Commission on Microbiological Specifications for Foods, Sampling for Microbiological Analysis: *Principles and Scientific Applications*, 2nd ed., Vol 2. University of Toronto Press, Toronto.

Jay, J. M., Loessner, J. L., and Golden, D. A. (2005). Chapter 2: Taxonomy role and significance of microorganism in foods. Chapter 13: Food protection with chemicals, and by biocontrol. In *Modern Food Microbiology* (7th ed.). J. M. Jay, J. L. Loessner, and D. A. Golden (Eds.). Springer Science + Business Media, Inc., New York.

Juliano, C., Mattana, A., and Usai, M. (2000). Composition and *in vitro* antimicrobial activity of the essential oil of Thymus herba-herbona Lasel growing wild in Sardinia. *Journal of Essential Oil Research* **12**, 516–522.

Kalchayand, N., Hanlin, M. B., and Ray, B. (1992). Sublethal injury makes Gram negative and gram positive bacteria sensitive to the bacteriocins, pediocin AcH and nisin. *Letters of Applied Microbiology* **15**, 239–243.

Kim, Y. H., Nam, K. C., and Ahn, D. U. (2002). Volatile profiles, lipid oxidation and sensory characteristics of irradiated meat from different animal species. *Meat Science* **61**, 257–265.

Labadie, J. (1999). Consequences of packaging on bacterial growth. Meat is an ecological niche. *Meat Science* **52**, 299–305.

Lambert, R. J. W., Skandamis, P. N., Coote, P., and Nychas, G. J. E. (2001). A study of the minimum inhibitory concentration and mode of action of oregano essential oil, thymol and carvacrol. *Journal of Applied Microbiology* **91**, 453–462.

Marino, M., Bersani, C., and Comi, G. (1999). Antimicrobial activity of the essential oils of Thymus vulgaris measured using a bioimpediometric method. *Journal of Food Protection* **62**(9), 1017–1023.

Mead and Adams (1977). Selective medium for rapid isolation of Pseudomonas associated with poultry meat spoilage. *British Poultry Science* **18**(6), 661–670.

Nilsson, L., Chen, Y., Chikindas, M. L., Huss, H. H., Gram, L., and Montville, T. J. (2000). Carbon dioxide and nisin act synergistically on *Listeria monocytogenes*. *Applied Environmental Microbiology* **66**, 769–774.

Ouattara, B., Simard, R. E., Holley, R. A., Piette, G. J. P., and Begin, A. (1997). Antibacterial activity of selected fatty acids and essential oils against six meat spoilage organisms. *International Journal of Food Microbiology* **37**, 155–162.

Pearson, D. (1991). Composition and analysis of foods. In *Longman ScientiWc & Technical*. R. Kirk and R. Sawyer (Eds.). London, pp. 642–643.

Racanicci, A. M. C., Danielsen, B., Menten, J. F. M., Regitano-d'Arce, M. A. B., and Skibsted, L. H. (2004). Antioxidant effect of dittany (*Origanum dictamnus*) in pre-cooked chicken meat balls during chill-storage in comparison to rosemary (*Rosmarius officinalis*). *European Food Research and Technology* **218**, 521–524.

Ratledge, C. and Wilkinson, S. G. (1988). An overview of microbial lipids. In *Microbial Lipids, Vol. 1*. C. Ratledge, and S. G. Wilkinson (Eds.). Academic Press, London, pp. 3–22.

Skandamis, P. and Nychas, G. J. E. (2001). Effect of oregano essential oil on microbiological and physico-chemical attributes of minced meat stored in air and modified atmospheres. *Journal of Applied Microbiology* **91**, 1011–1022.

Skandamis, P., Tsigarida, E., and Nychas, G. J. E. (2002). The effect of oregano essential oil on survival/death of Salmonella typhimurium in meat stored at 5°C under aerobic VP/MAP conditions. *Food Microbioogy* **19**, 97–103.

Steel, R. G. D. and Torrie, J. H. (1980). Principles and Procedures of Statistics. *A Biometrical Approach*. Mc Graw-Hill, New York.

Stevens, K. A., Sheldon, B. W., Klapes, N. A., and Klaenhammer, T. R. (1991). Nisin treatment for the inactivation of *Salmonella* species and other gram negative bacteria. *Applied Environmental Microbiology* **57**, 3613–3615.

Szabo, E. A. and Cahill, M. E. (1998). The combined effects of modified atmosphere, temperature, nisin and ALTA 2341 on the growth of *Listeria monocytogenes*. *International Journal of Food Microbiology* **43**, 21–31.

Thomas, L. V., Clarkson, M. R., and Delves-Broughton, J. (2000). Nisin. In *Natural Food*

Antimicrobial Systems. A. S. Naidu (Ed.). CRC Press, USA, pp. 463–524.

US Food and Drug Administration, (1988). Nisin Preparation: Affirmation of GRAS status as direct human food ingredient. *Federal Register* **53**, April 6.

Vaara, M., (1992). Agents that increase the permeability of the outer membrane. *Microbiology Reviews* **56**(3), 395–411.

Zaika, L. L., Kissinger, J. G., and Wasserman, A. E., (1983). Inhibition of lactic acid bacteria by herbs. *Journal of Food Science* **48**, 1455–1459.

Zeitoun, A. A. M., Debevere, J. M., and Mossel, D. A. A. (1994). Significance of Enterobacteriaceae as index organisms for hygiene of fresh untreated poultry, poultry treated with lactic acid and poultry stored in modified atmosphere. *Food Microbiology* **11**, 169–176.

8

Adam, K., Sivropoulou, A., Kokkini, S., Lanaras, T., and Arsenakis, M., (1998). Antifungal activities of *Origanum vulgare* subsp. *hirtum*, *Mentha spicata*, *Lavandula angustifolia* and *Salvia fruticosa* essential oils against human pathogenic fungi. *Journal of Agricultural and Food Chemistry* **46**, 1739–1745.

Aureli, P., Constantini, A., and Zolea S. (1992). Antimicrobial activity of some plant essential oils against *Listeria monocytogenes*. *Journal of Food Protection* **55**, 344–348.

Botsoglou, N. A., Grigoropoulou, S. M., Botsoglou, E., Govaris, A., and Papageorgiou, G., (2003). The effects of dietary oregano essential oil and a-tocopheryl acetate on lipid oxidation in raw and cooked turkey during refrigerated storage. *Meat Science* **65**, 1193–1200.

Burt, S., (2004). Essential oils: Their antibacterial properties and potential applications in foods-a review in press. *International Journal of Food Microbiology* **94**(3), 223–253.

Chouliara I. and Kontominas M. G., (2006). Combined effect of thyme essential oil and modified atmosphere packaging to extend shelf life of fresh chicken meat. In *Recent Progress in Medicinal Plants: Natural Product (15)*. J. N. Govil, V. K. Singh, K. Almad, and R. K. Sharma (Eds.)..Studium Press, LLC, USA, pp. 423-442.

Chouliara, I., Savvaidis, I., Riganakos, K., and Kontominas, M. G., (2005). Shelf life extension of vacuum-packaged sea bream (*Sparus aurata*) fillets by combined γ-irradiation and refrigeration: microbiological, chemical and sensory changes. *Journal Science Food Agriculture* **85**, 779–784.

Chouliara, E., Karatapanis, A., Savvaidis, I. N., and Kontominas, M. G. (2007). Combined effect of oregano essential oil and modified atmosphere packaging on shelf life extension of fresh chicken breast meat, stored at 4°C. *Food Microbiology* **24**, 607–617.

Chung, K. T., Dickson, J. S., and Crouse, J. D., (1989). Effects of nisin on growth of bacteria attached to meat. *Applied Environmental Microbiology* **55**,1329–1333.

Clarkson, and Delves-Broughton, J. (2000) In *Natural Food Antimicrobial Systems*. A. S. Naidu, L. V. Thomas, and M. R. Nisin (Eds.). CRC Press, USA, pp. 463–524.

Davies, E. A., Bevis, H. E., and Delves-Broughton, J., (1997). The use of the bacteriocin, nisin, as a preservative in ricotta-type cheeses to control the food-borne pathogen *Listeria monocytogenes*, *Letters of Applied Microbiology* **24**, pp. 343–346.

Deans, S. G. and Richie, G., (1987). Antimicrobial properties of plant essential oils. *International Journal of Food Microbiology* **5**, 165–180.

Devlieghere, F., Vermeiren, L., and Debevere, J., (2004). New preservation technologies: Possibilities and limitations. *International Dairy Journal* **14**, 273–285.

de Vuyst, L., and Vandamme, E. (1994). Bacteriocins of Lactic Acid Bacteria. *Microbiology, Genetics and Applications*, Blackie Academic and Professional, London.

Fang, T. J., Lin, L. W. (1994). Growth of *Listeria monocytogenes* and *Pseudomonas fra*gi on cooked pork in a modified atmosphere packaging/nisin combination. *Journal of Food Protection* **57**, 479–485.

FAO/WHO Working Group on Risk Assessment of Microbiological Hazards in Foods, 2002. Preliminary Report—Hazard Identification, Hazard Characterization and Exposure Assessment of *Campylobacter* spp. in Broiler Chickens (MRA 01/05).

Farkas, J., (1990). Combination of irradiation with mild heat treatment, *Food Control* **1**, 223–229.

Ferreira, M. A. and Lund, B. M., (1996). The effect of nisin on *Listeria monocytogenes* in culture medium and long-life cottage cheese. *Letters of Applied Microbiology* **22**, 433–438.

Helander, I. M., Alakomi, H. L., Latva-Kala, K., Mattila-Sandholm, T., Pol, I., Eddy, J. Smid, E. J., Gorris, L. G. M., and Wright, A. (1998). Characterization of the Action of Selected Essential Oil Components on gram-negative bacteria. *Journal of Agriculture and Food Chemistry* **46**, 3590–3595.

Jay, J. M. (1986) Microbial spoilage indicators and metabolites. In *Foodborne Microorganisms and their toxins. Developing methodology.* M. D. Pierson and A. Stern (Eds.), Marcel Dekker Inc. Basel, pp. 213–240.

Jay, J. M., (2000). *Modern Food Microbiology*, 6th ed. Aspen Publishers, Inc., Gaithersburg, MD, p. 679.

Juliano, C., Mattana, A., and Usai, M., (2000). Composition and *in vitro* antimicrobial activity of the essential oil of Thymus herba-herbona Lasel growing wild in Sardinia. *Journal of Essential Oil Research* **12**, 516–522.

Kalchayand, N., Hanlin, M. B., and Ray, B. (1992). Sublethal injury makes gram-negative and gram-positive bacteria sensitive to the bacteriocins, pediocin AcH and nisin. *Letters of Applied Microbiology* **15**, 239–243.

Karabagias, I., Badeka, A. V., and Kontominas, M. G., (2010). Shelf-extension of lamb meat using thyme or oregano essential oils and modified atmosphere packaging. *Meat Science* **88**(1), 109–116.

Kim, J., Marshall, M. R., and Wei, C. I., (1995). Antibacterial activity of some essential oil components against five foodborne pathogens. *Journal of Agriculture and Food Chemistry* **43**, 2839–2845.

Lambert, R. J. W., Skandamis, P. N, Coote, P., and Nychas, G. J. E. (2001). A study of the minimum inhibitory concentration and mode of action of oregano essential oil, thymol and carvacrol. *Journal of Applied Microbiology* **91**, 453–462.

Marino, M., Bersani, C., and Comi G., (1999). Antimicrobial activity of the essential oils of *Thymus vulgaris* measured using a bioimpediometric method. *Journal of Food Protection* **62**(9), 1017–1023.

Montville, T. J. and Chen, Y., (1998). Mechanistic action of pediocin and nisin: Recent progress and unresolved questions. *Applied Microbiology and Biotechnology* **50**, 511–519.

Mytle, N., Anderson, G. L., Doyle, M. P., and Smith, M. A. (2006). Antimicrobial activity of clove (*Syzgium aromaticum*) oil in inhibiting *Listeria monocytogenes* on chicken frankfurters. *Food Control* **17**(2), 102–107.

Pawar, D. D., Malik, S. V. S., Bhilegaonkar, K. N., and Barbuddhe, S. B., (2000). Effect of nisin and its combination with sodium chloride on the survival of *Listeria monocytogenes* added to raw buffalo meat mince. *Meat Science* **56**, 215–219.

Shelef, L. A., Naglik, O. A., and Bogen, D. W. (1980). Sensitivity of some common food-borne bacteria to the spices sage, rosemary and all spice. *Journal of Food Science* **45**, 1042–1044.

Sivropoulou, A., Papanikolaou E., Nikolaou, K., Kokkini, J., Loukatos, T., and Arsenakis A (1996). Antimicrobial and cytotoxic activities of oreganum essential oils. *Journal of Agriculture and Food Chemistry* **44**, 1202–1205.

Skandamis, P. and Nychas, G. J. E., (2001). Effect of oregano essential oil on microbiological and physico-chemical attributes of minced meat stored in air and modified atmospheres. *Journal of Applied Microbiology* **91**, 1011–1022.

Skandamis, P., Tsigarida, E., and Nychas, G. J. E. (2002). The effect of oregano essential oil on survival/death of *Salmonella typhimurium* in meat stored at 5°C under aerobic VP/MAP conditions. *Food Microbiology* **19**, 97–103.

Solomakos, N., Govaris, A., Koidis, P., and Botsoglou, N. (2008). The antimicrobial effect of thyme essential oil, nisin, and their combination against *Listeria monocytogenes* in minced beef during refrigerated storage. *Food Microbiology* **25**, 120–127.

Stecchini, M. L., Sarais, I., and Giavedoni, P. (1993). Effect of essential oils on *Aeromonas hydrophila* in a culture medium and in cooked pork. *Journal of Food Protection* **56**, 406–409.

Steel, R. G. D. and Torrie, J. H. (1980). Principles and Procedures of Statistics. *A Biometrical Approach*. Mc Graw-Hill, New York.

Stevens, K. A., Sheldon, B. W., Klapes, N. A., and Klaenhammer, T. R. (1991). Nisin treatment for the inactivation of *Salmonella* species and other gram-negative bacteria. *Applied Environmental Microbiology* **57**, 3613–3615.

Tsigarida, E., Skandamis, P., and Nychas, G. J. E., (2000). Behavior of *Listeria monocytogenes* and autochthonous flora on meat stored under aerobic, vacuum and modified atmosphere packaging conditions with or without the presence of oregano essential oil at 5°C. *Journal of Applied Microbiology* **89**, 901–909.

Vingolo, G., Palacios, J., Farias, M. E., Sesma, F., Schillinger, U., Holzapfel, W., and Oliver, G. (2000). Combined effect of bacteriocins on the survival of various *Listeria* species in broth and meat system. *Current Microbiology* **41**, 410–416.

Yuste, J., Mor-Mur, M., Capellas, M., Guamis, B., and Pla, R. (1998). Microbiological quality of mechanically recovered poultry meat treated with high hydrostatic pressure and nisin. *Food Microbiology* **15**, 407–414.

Zaika, L. L., Kissinger, J. G., and Wasserman, A. E., (1983). Inhibition of lactic acid bacteria by herbs. *Journal of Food Science* **48**, 1455–1459.

9

Berenzon, S. and Saguy, I. S. (1998). Oxygen absorbers for extension of crackers shelf life. *LWT Food Science and Technology* **31**, 1–5.

Chouliara, E., Karatapanis, A., Savvaidis, I. N., and Kontominas, M. G. (2007). Combined effect of oregano essential oil and modified atmosphere packaging on shelf life extension of fresh chicken breast meat, stored at 4°C. *Food Microbiology* **24**, 607–617.

Darmadji, P. and Izumimoto, M. (1994). Effect of chitosan in meat preservation. *Meat Science* **38**, 243–254.

Day, B. P. F. (1989). Extension of shelf life of chilled foods. *European Food & Drink Reviews* **4**, 47–56.

Dawson, P. L., Hon, H., Vollet, L. M., Clardy, L. B., Martinez, R. M., and Acton, J. C. (1995). Film oxygen transmission rate effects on ground chicken meat quality. *Poultry Science* **14**, 1381–1387.

Du, M., Hur, S. J., and Ahn, D. U., (2002). Raw meat packaging and storage affect the color and odor of irradiated breast fillets after cooking. *Meat Science* **61**, 49–54.

Georgantelis, D., Ambrosiadis, I., Katikou, P., Blekas, G., and Georgakis, S. A. (2007). Effect of rosemary extract, chitosan and α-tocopherol on microbiological parameters and lipid oxidation of fresh pork sausages stored at 4°C. *Meat Science* **76**, 172–181.

Giatrakou, V., Ntzimani, A., and Savvaidis, I. N., (2010). Effect of chitosan and thyme oil on a ready to cook chicken product. *Food Microbiology* **27**, 132–136.

Giatrakou, V., Ntzimani, A., and Savvaidis, I. N., (2010). Combined chitosan-thyme treatments with modified atmosphere packaging on a ready-to-cook poultry product. *Journal of Food Protection* **73**(4), 663–669.

Helander, I. M., Nurmiaho-Lassila, E. L., Ahvenainen, R., Rhoades, J., and Roller, S. (2001). Chitosan disrupts the barrier properties of the outer membrane of gram-negative bacteria. *International Journal of Food Microbiology* **71**, 235–244.

ICMSF (1986). International Commission on Microbiological Specifications for Foods. *Sampling for Microbiological Analysis: Principles and Scientific Applications, 2nd ed., Vol 2*. University of Toronto Press, Toronto.

Jay, J. M., Loessner, J. L., and Golden, D. A. (2005). Chapter 2: Taxonomy role and significance of microorganism in foods. Chapter. 13: Food protection with chemicals and biocontrol. In *Modern Food Microbiology* (7th ed.). J. M. Jay, J. L. Loessner, and D. A. Golden (Eds.). Sringer Science & Business Media, Inc., New York.

Kirk, R. S. and Sawyer, R. (1991). In *Composition and Analysis of Foods, Longman Scientific & Technical (9 th ed.)*. Pearson, Essex CM20 2JE, England.

Labuza, T. P. (1987). Oxygen scavenger sachets. *Food Research International* **32**, 276–277.

Lin, K. W. and Chao, J. Y., (2001). Quality characteristics of reduced-fat Chinese-style sausage as related to chitosan's molecular weight. *Meat Science* **59**, 343–351.

Martinez, L., Djenane, D., Cilla, I., Beltran, J. A., and Roncales, P. (2006). Effect of varying oxygen concentrations on the shelf life of fresh pork sausages packaged in modified atmosphere. *Food Chemistry* **94**, 219–225.

Mohan, C. O., Ravishankar, C. N., and Srinivasagopal, K. (2008). Effect of O_2 scavenger on the shelf life of catfish (*Pangasius sutchi*) steaks during chilled storage. *Journal of science and Food Agriculture* **88**, 442–448.

Nakamura, H. and Hoshino, J. (1983). *Techniques for the preservation of food and employment of an oxygen absorber in technical information*. Ageless Division. Mitsubishi Gas Chemical Co., Tokyo, pp. 1–45.

Ouattara, B., Simard, R. E., Holley, R. A., Piette, G. J. P., and Begin, A. (1997). Antibacterial activity of selected fatty acids and essential oils against six meat spoilage organisms. *International. Journal. of Food Microbiology* **37**, 155–162.

Rudrapatnam, N. T. and Farooqahmed, S. K. (2003). Chitin-the undisputed biomolecule of great potential. *Critical Reviews in Food Science and Nutrition* **43**, 61–87.

Sagoo, S., Board, R., and Roller, S. (2002). Chitosan inhibits growth of spoilage microorganism in chilled pork products. *Food Microbiology* **19**, 175–182.

Shahidi, F., Arachchi, J. K. V., and Jeon, Y. J. (1999). Food applications of chitin and chitosans. *Trends in Food Science and Technology* **10**, 37–51.

Sheridan, J. J., Doherty, A. M., Allen, P., McDowell, D. A., Blair, I. S., and Harrington, D. (1997). The effect of vacuum and modified atmosphere packaging on the shelf life of lamb primal stored at different temperatures. *Meat Science* **45**, 107–117.

Soultos, N., Tzikas, Z., Abrahim, A., Georgantelis, D., and Ambrosiadis, I. (2008). Chitosan effects on quality properties of Greek style fresh pork sausages. *Meat Science* **80**, 1150–1156.

Steel, R. G. D. and Torrie, J. H. (1980). Principles and Procedures of Statistics. *A Biometrical Approach*. Mc Graw-Hill, New York.

Yen, M. T., Yang, J. H., and Mau, J. L. (2008). Antioxidant properties of chitosan from crab shells. *Carbohydrate Polymers* **74**, 840–844.

Zeitoun, A. A. M., Debevere, J. M., and Mossel, D. A. A. (1994). Significance of Enterobacteriaceae as index organisms for hygiene of fresh untreated poultry, poultry treated with lactic acid and poultry stored in modified atmosphere. *Food Microbiology* **11**, 169–176.

Zheng, L. Y. and Zhu, J. F., (2003). Study on antimicrobial activity of chitosan with different molecular weights. *Carbohydrate Polymers* **54**, 527–530.

10

Al-Habbal, M. J., Al-Habbul, Z., Huwez, and F.U. (1984). A double-blind controlled clinical trial of mastic and placebo in the treatment of duodenal ulcer. *Journal of Clinical and Experimental Pharmacology and Physiology* **11**(5), 541–544.

Al-Said, M. S., Ageel, A. M., Parmar, N. S., and Tariq, M. (1986). Evaluation of mastic, a crude drug obtained from *Pistacia lentiscus* for gastric and duodenal anti-ulcer activity. *Journal of Ethnopharmacology* **15**(3), 271–278.

Anonymous (2001). Innovation is a cultural thing. *Dairy Foods* **102**, 12–13.

Besser, R. E., Lett, S. M., Weber, J. T., Doyle, M. P., Barret, T. J., Wells, J. G., and Griffin, P. M. (1993). An outbreak of diarrhea and haemolytic uremic syndrome from *E.coli* O157:H7 in fresh pressed apple cider. *Journal of the American Medical Association* **22**(69), 2217–2220.

Burt, S. (2004). Essential oils: Their antibacterial properties and potential applications in foods. *International Journal of Food Microbiology* **94**(3), 223–253.

Canganella, F., Ovidi, M., Paganini, S., Vettraino, A. M., Bevillacqua, L., and Trovatelli, L. D. (1998). Survival of undesirable micro-organisms in fruit yoghurts during storage at different temperatures. *Food Microbiology* **15**, 71–77.

Cosentino, S., Tuberoso, C. A., Pisano, B., Sattla, M., Mascia, V., Arzedi, E., and Palmas, F. (1999). *In vitro* antimicrobial activity and chemical composition of Sardinian Thymus essential oils. *Letters of Applied Microbiology* **29**, 130–135.

Con, A. H., Cakmakci, Caglar, A., and Gokalp, H. Y. (1996). Effects of different fruits and storage period on microbiological qualities of fruit flavored yogurt produced in Turkey. *Journal of Food Protection* **59**(4), 402–406.

Davidson, P. M. (1997). Chemical preservatives and natural antimicrobial compounds. In *Food Microbiology Fundamentals and Frontiers*. M. G. Doyle, L R Beuchat, and T. J. Montville (Eds.). ASM Publications, Washington DC, pp. 520–556.

Deak, T. (1991). Foodborne yeasts. *Advances in Applied Microbiology* **36**, 179–278.

Desmarchelier, P. and Grau, F. H. (1997). Escherichia coli. In *Foodborne Microorganisms of Public Health Significance*. 5th ed. A. O. Hocking, G. Arnold, I. Jenson, K. Newton, and P. Sutherland (Eds.). Australian Institute of Food Science and Technology, Sydney, Australia, pp. 231–263.

Doukas, C. (2003). Cosmetics that contain mastic gum and mastic oil. *Chemistry Chronical* **12**, 36–39.

El-Nawawy, M. A., El-Kenany, Y. M., and El-Chaffau, E. A. (1998). Effect of herb plants on the use of yogurt culture. *Annuals of Agricultural. Society*, 7th Conference Agriculture Development Research. Faculty of Agriculture, Alin Shams University of Cairo, Egypt, 15–17 December.

Errendilek, G. A. (2007). Survival of *Escherichia coli* O157:H7 in yogurt drink, plain yogurt and salted (tuzlu) yogurt: Effects of storage time, temperature, background flora and product characteristics. *International Journal of Dairy Technology* **60**, No 2 May, 2007.

Hamann, W. T. and Marth, E. H. (1984). Effects of cysteine and different incubation temperatures on the microflora, chemical composition and sensory characteristics of bio-yogurt made from goat's milk. *Food Chemistry* **100**, 788–793.

Harpaz, S., Glatman, L., Drebkin, V., and Gelman, A. (2003). Effects of herbal essential oils used to extend the shelf life of freshwater Asian sea bass fish (Lates calcarifer) *Journal of Food Protection* **66**(3), 410–417.

Hassan, M. V. A., El-Nassar, M. A., and Abou Dawood, S. A. (2001). Antimycotic and antimycotoxigenic activity of some spices and herbs. *Proceeding 8th: Egyptian Conference of Dairy Technology*, pp. 603–609.

Hsin-Yi, C. and Chou, C. C. (2001). Acid adaptation and temperature effect on the survival of *E. coli* O157:H7 in acidic fruit juice and lactic fermented milk product. *Intrernational Journal of Food Microbiology* **70**, 189–195.

http://chimikoergastirio.blogspot.com

http://en.wikipedia.org/wiki/Almond

http://www.raysahelian.com/almond.html

Huwez, F. U. (1998) Mastic gum kills *Helicobacter pylori. New England Journal Medicine* **339**, 1946–1949.

Iauk, L., Ragusa, S., Rapisadra, A., France, S., and Nicolosi, V. M. (1996). *In vitro* antimicrobial activity of *Pistachio lentiscus* L. extracts: Preliminary report. *Journal of Chromatography* **8**(3), 207–209.

Jaziri, I., Ben Slama, M., Mhadhbi, H., Urdaci, M., and Hamdi, M. (2009). Effect of green and black teas on the characteristic microflora of yogurt during fermentation and refrigerated storage. *Food Chemistry* **42**, 614–620.

Jeon, H. J., Lee, K. S., and Ahn, Y. J. (2001). Growth-inhibiting effects of constituents of *Pinus densiflora* leaves on human intestinal bacteria. *Food Science and Biotechnology* **10**(4), 403–407.

Kang, D. U. and Fung, D. Y. C. (1999). Effect of diacetyl on controlling *E. coli* O157:H7 and *Salmonella typhimurium* in the presence of starter culture in a laboratory medium and during meat fermentation. *Journal of Food Protection* **62**(9), 975–979.

Kasimoglou, A. and Akgun, J. (2004). Survival of *E. coli* O157:H7 in the processing and post-processing stages of acidophilus yogurt. *International Journal of Food Science and Technology* **39**, 563–568.

Kim, F., Marshall, M. R., and Wei, C. (1997). Antibacterial activity of some essential oil components against five foodborne pathogens. *Journal of Agricultural and Food Chemistry* **43**, 2839–2845.

Kotzekidou, P., Giannakidis, P., and Boulamatsis, A. (2008). Antimicrobial activity of some plant extracts and essential oils against foodborne pathogens *in vitro* and on the fate of inoculated pathogens in chocolate. *London Weekend Television* **41**, 119–127.

Koutsoudaki, C., Krjek, A., and Rodger, A. (2005). Chemical Composition and Antibacterial Activity of the Essential Oil and the Gum of *Pistacia lentiscus Var. Chia. Agricultural and Food Chemistry* **53**, 7681–7685.

Lambert, R. J. W., Skandamis, P. N., Coote, P., and Nychas, G. J. E. (2001). A study of the minimum inhibitory concentration and mode of action of oregano essential oil, thymol and carvacrol. *Journal of Applied Microbiology* **91**, 453–462.

Lee, S. M. and Chen, J. (2005). The influence of an extracellular polysaccharide, comprised of colonic acid, on the fate of *E. coli* O157:H7 during processing and storage of stirred yogurt. *London Weekend Television* **38**,785–790.

Leyer, G. J., Wang, L., and Johnson, E. A. (1995). Acid adaptation of *Escherichia coli* O157:H7 increase survival in acidic foods. *Applied and Environmental Microbiology* **61**, 3752–3755.

Magiatis, P., Melliou, E., Skaltsounis, A. L., Chinou, I. B., and Mitakou, S. (1999). Chemical composition and Antimicrobial activity of the essential oils of *Pistacia lentiscus* var. *chia*. *Planta Medica* **65**, 749–752.

Massas, S., Altier, C., Quaranta, V., and de ,Pace. R. (1997). Survival of *E.coli* O157:H7 in yoghurt during preparation and storage at 4°C. *Letters of Applied Microbiology* **24**, 342–350.

Morgan, P., Newman, C. P., Hutchinson, D. N., Walker, A. M., Rowe, B., and Majid, D. E. (1993). Verotoxin producing *Escherichia coli* 0157 infections associated with the consumption of yoghurt. *Epidemiology and Infection* **111**, 181–187.

Neagan, S. D., Bryant, J. L., and Bark, D. H. (1994). Survival of *E.coli* O157:H7 in mayonnaise and mayonnaise-based sauce at room and refrigerated temperatures. *Journal of Food Protection* **57**, 629–631.

Nikaido, H. (1996). Outer membrane. In *Escherichia coli and Salmonella: Cellular and molecular biology*. F. C. Neidhardt (Ed.). Vol 1, ASM Press, Washington AC, pp. 29–47.

Ogwaro, B. A., Gibson, H., Whitehead, M., and Hill, D. J. (2002). Survival of *Escherichia coli O157:H7* in traditional African yoghurt fermentation. *International Journal of Food Microbiology* **79**, 105–112.

Otaibi, M. A. and Demerdash, E. (2008). Improvement of quality and shelf life of concentrated yogurt (Labneh) by the addition of some essential oils. *African Journal of Microbiology* **2**, 156–161.

Penney, V., Henderson, G., Blurnb, C., and Johnson-Green, P. (2004). The potential of phytopreservatives and nisin to control microbial spoilage of minimally processed fruit yogurts. *Innovative Food Science and Emerging Technologies* **5**, 369–375.

Ryu, J. H. and Beuchat, L. (1998). Influence of acid tolerance responses on survival, growth and thermal cross-protection of *Escherichia coli* O157:H7 in acidified media and fruit juices. *International Journal of Food Microbiology* **45**, 185–193.

Schelz, Z. and Molnar-Hohaman, J. (2006). Antimicrobial and antiplasmid activities of essential oils. *Fitotherapia* **77**(4), 279–285.

Sikkema, J., de Bont, J. A. M., and Poolman, B. (1995). Mechanism of membrane toxicity of hydrocarbons. *Microbiology Reviews* **59**, 201–222.

Tin, L. Z., Ho, Y. W., Abdullah, N., Ali, M., and Jalaludin, S. (1998). Antagonistic effects of intestinal Lactobacillus isolates on pathogens of chicken. *Letters of Applied Microbiology* **23**, 67–71.

Topitsoglou-Themeli, V., Dagalis, P., and Lambrou, D. A. (1984). Chios mastic chewing gum and oral hygiene. I. The possibility of reducing or preventing microbial plaque formation. *Hellenika Stomatologika Chronika* **28**(3), 166–170.

Varnam, A. H. and Sutherland, J. P. (1996) *Milk and milk products*. Chapman and Hall, London, chapter.1, 2, 8.

Vinderola, C. G., Costa, G. A., Regenhardt, S., and Reinheimer, J. A. (2002). Influence of compounds associated with fermented dairy products on the growth of Lactic acid starter and probiotic bacteria. *International Dairy Journal* **12**, 579–589.

Index

M

N

For Product Safety Concerns and Information please contact our EU
representative GPSR@taylorandfrancis.com
Taylor & Francis Verlag GmbH, Kaufingerstraße 24, 80331 München, Germany

www.ingramcontent.com/pod-product-compliance
Ingram Content Group UK Ltd.
Pitfield, Milton Keynes, MK11 3LW, UK
UKHW020202200926
475702UK00004B/579

* 9 7 8 1 9 2 6 8 9 5 0 7 9 *